LIFE: ORIGIN AND EVOLUTION

Readings from
**SCIENTIFIC
AMERICAN**

LIFE: ORIGIN AND EVOLUTION

With Introductions by
Clair Edwin Folsome
University of Hawaii at Manoa

W. H. Freeman and Company
San Francisco

Dedicated to Jo

Most of the *Scientific American* articles in *Life: Origin and Evolution* are available as separate Offprints. For a complete list of more than 1200 articles now available as Offprints, write to W. H. Freeman and Company, 660 Market Street, San Francisco, California 94104.

Library of Congress Cataloging in Publication Data

Main entry under title:

Life: origin and evolution.

   Bibliography: p.
   Includes index.
   1. Life—Origin—Addresses, essays, lectures.
I. Folsome, Clair Edwin, 1935–   II. Scientific American.
QH325.L693 1978   577   78–15129
ISBN 0–7167–1033–1
ISBN 0–7167–1032–3 pbk.

PREFACE

This collection provides a short series of articles and interpretations that focus on the origin of life on our planet. Their points of view range from astronomy to the evolution of proteins. The book can complement texts and courses in fields from astronomy to geology, or it can be read alone as an introduction to this fascinating field. Introductions to the various sections attempt to map out the tortuous pathways of current research and to point out the central themes of theories on the origin of life.

Controversies abound in this arena of little facts and vast conjectures about a 4000-million-year past. I have attempted to be as consistent as possible within the limits of presently known facts. Great voids in our knowledge remain. Significant recent advances in the planetary sciences, astronomy, and cosmology allow us to reappraise the original origin-of-life theories of A. I. Oparin and J. D. Bernal in light of the current facts of planetary astronomy. Although the details of the Oparin–Bernal scenario were incorrect, its central ideas remain sound.

Two major problems face present-day workers in this field. First, how can polymers such as peptides and polynucleotides be formed in a dilute aqueous environment? And second, how did a mechanism as elaborate and reflexive as the genetic apparatus arise? The current research literature abounds with suggestions, which will be outlined in later introductions. At present, however, none of these conjectures has compelling experimental support.

Many of the suppositions made in these articles could truly be tested if we were able to study even one example of extraterrestrial life. This eventuality does not appear possible, since we are alone as a life form within our solar system. This aspect of reality forces us to retreat to the laboratory or to seek communication with intelligences outside the solar system. The laboratory, if constrained within realistic boundaries of simulation and experimental design and supported by imagination, can generate data to permit us to develop our hypotheses further. The advances thus made should enable us to judge sensibly how probable our existence is within the Universe.

Communication with extraterrestrial intelligences depends ultimately upon the accuracy of our laboratory work. We can try to communicate out of desperation, out of faith, or from the near certainty that someone else must be there. Let's hope that we can show by experiment that someone else might trade thoughts with us.

Kailua, Hawaii   Clair Edwin Folsome
June, 1978

Note on cross-references to SCIENTIFIC AMERICAN *articles:* Articles included in this book are referred to by title and page number; articles not included in this book but available as Offprints are referred to by title and offprint number; articles not included in this book and not available as Offprints are referred to by title and date of publication.

CONTENTS

LIFE: ORIGIN AND EVOLUTION

GENERAL INTRODUCTION

General Introduction

This collection of articles is designed to supplement textbooks on astronomy, geology, biology, and the origin of life. It also points out what our current major uncertainties are and what future research work might most profitably be done. Rarely has the origin-of-life question itself been the subject of a college course. This fascinating and elusive topic inevitably appears during introductory work in astronomy, geology, biology, and other fields of science and philosophy. Our local cosmogony now rests upon a series of postulates, each of which might be a correct representation of reality. But when we combine all the postulates and sum their error factors, we face quite an uncertain picture. When thoughtfully combined, these assumptions give us a mental picture of our origin and a philosophical workspace from which to appreciate and extend our view—be it of the Universe or a single subatomic particle.

A. I. Oparin and J. D. Bernal were among the first contemporary scientists to attempt explanations of life's origins. They did so, of necessity, within the constructs of the sciences of their time. Since then, our ideas of the Universe and of our world have changed drastically. We retain the wondrous heritage of their central ideas, which continue to bear fruit for this field of study. For example, Oparin thought that the primitive Earth had an atmosphere of methane, ammonia, and hydrogen. This view, still held by many today, was based on the then recently discovered cosmic abundances of the elements, which showed clearly that hydrogen was far in excess of all the other elements. Thus the dominant forms of carbon and nitrogen would necessarily have been reduced to methane and ammonia. Oparin was both right and wrong. We know now that cosmic abundances of the elements do not apply to the terrestrial planets, which have been stripped almost entirely of the less massive volatile elements because of the high early temperatures of the forming protosun. However, he was also right because any atmosphere formed by an evolving proto-Earth would contain some hydrogen and other entrained volatiles, but no oxygen.

Oparin and his contemporaries conceived of the early terrestrial environment as a Darwinian "warm little pond," teeming with a multitude of organic molecules formed by successive reactions of methane, ammonia, water, and phosphate. This scenario of the primitive Earth leads one to conceive of pre-life oceans as rich in organic molecules as a gourmet chicken soup. In truth, the primitive oceans contained extraordinarily small amounts of organic compounds. Why the disparity? Because we must consider not only the formation rates of organic compounds, but their rates of destruction as well. Recent calculations by Klaus Dose show that any single amino acid would be present in steady-state amounts of some 10^{-7} moles—quite dilute indeed, a far cry even from commercial chicken soup. How could life arise in such a dilute environment?

The Oparin hypothesis, based on high concentrations of organic compounds, goes on to assume that polymers form. Proteins, nucleic acids, and polysaccharides become abundant and, as they must, interact to generate random but structured groupings called colloids. As a matter of fact, colloidal particles are simple to make. Mix concentrated solutions of histone (a protein) and gum arabic (a polysaccharide). Adjust the pH. Suddenly the once transparent solution of macromolecules becomes cloudy and turbid. Under the microscope, cell-like and cell-size particles are visible. These colloidal associations of macromolecules can mimic many properties of truly biological cells. From these observations Oparin inferred that early protocells arose by a long evolution from colloidal systems. Continued evolution from protocells to cells such as bacteria completes an explanation for the origin of cellular life. But— and this is the crux of the issue—where did the extremely concentrated macromolecular solution come from in the first place? From a dilute soup it is most difficult to synthesize even small polymers. Those that might be made have a higher probability of being destroyed than of interacting with one another.

It is possible to evade the "concentration gap" by calling upon unique and unusual environmental changes that appear likely until examined critically. For instance, let us evaporate most of the water from a shallow lake that contains dilute small organic molecules. As the pond approaches dryness, the concentration of all organic species greatly increases, and their interaction to form a polymer seems inevitable. But this is a wish. Not only do all organic molecules become concentrated in this environment, but inorganic salts— sodium chloride and others—are similarly concentrated. Since the latter far outnumber the former, the chance of a profitable encounter between organic molecules to form polymers is vanishingly small. In addition, as lakes and ponds evaporate, the water depth decreases and solar ultraviolet radiation acts ever more strongly to decompose whatever small organic molecules remain.

We are faced with a serious problem. How did local high concentrations of small organic molecules, and their polymers, come to be?

Bernal approached this issue and proposed an alternative concentrative mechanism: clays. Clays are layered lattice silicates that contain sheet upon sheet of siliceous matter ideal for trapping and concentrating small organic compounds. When localized within clays, organic molecules can polymerize; their polymers can similarly be held by the clays.

Current knowledge seems to point in directions other than clays, although active research continues. One difficulty with clays is that they form from the weathering and sedimentary processes of stable ocean systems. Microfossils of the oldest cells are of the same age as the oldest sedimentary rocks—formed before clays became a major feature of the evolving terrestrial environment. Another problem is that clays adsorb those organic molecules required for a biology and remove them from phase-bounded (cell-like) systems. Hence clays, even if present, might seem to inhibit rather than promote the origin of life.

Every time we examine the specifics of the theories presented by Oparin and Bernal, current information seems to contradict them. Yet, in the main, they were right. How can one be both right and wrong? Easy—we do it all the time. Any explanation of a complex issue invariably has two components: the truth and the details. Oparin and Bernal were deficient in their details because, at that time, astronomy and geology had barely any to supply. But their truth still stands. They visualized life as originating from unorganized matter as a natural outcome of the forces that shape and evolve terrestrial planets. Their models were wrong, but the central theme they pursued seems even more right now than before.

We now have considerably more details. We have direct knowledge of the Moon, Mars, Venus, and Mercury. We now see the Earth with new eyes— as an ever-changing admixture of floating continents upon a denser, jumbled mantle and core. We also have a wealth of information on the evolution of proteins and cells, on the low likelihood of life on other planets in the solar system, and on the forces that led to the formation of our solar system.

The articles collected here show how the details of the Oparin–Bernal hypothesis have changed and developed, and how the truth conceived by these thinkers remains for us to expand into a truly coherent theory for the origin of life.

No theory encompassing as vast a range of time and events as this can ever be "proven" without evidence from critical experiments. One such critical experiment was NASA's *Viking 1* mission which searched for evidence of life on Mars. Such evidence was not found. Just a single example of extraterrestrial life would allow us to test all our detailed structures of hypotheses. Must proteins inevitably contain L-alpha amino acids, and must they serve as catalysts? Is DNA/RNA the exclusive genetic and protein-synthesizing system for all biologies? Must a biology have cells as its basic structural unit? These questions remain for our descendants, who will go to the stars to test our thoughts.

Suggested Further Reading

Bernal, J. D. 1967. *Origin of Life*. World Publishing Company, Ohio.

Folsome, Clair Edwin. 1979. *The Origin of Life: A Warm Little Pond*. W. H. Freeman and Company, San Francisco.

Kenyon, D. H. and G. Steinman. 1969. *Biochemical Predistination*. McGraw-Hill, New York.

Oparin, A. I. 1936. *The Origin of Life*. Macmillan, London.

I

FORMATION AND EARLY EVOLUTION OF THE EARTH

I   FORMATION AND EARLY EVOLUTION OF THE EARTH

INTRODUCTION

Speculation on the origin of life is futile unless it is based on some degree of knowledge about the environmental parameters at the time. Is our planet a rare phenomenon—a product of a series of events likely never to be repeated—or are planets a common result of the formation of stars? The articles by A. G. W. Cameron and Harold C. Urey implicitly adopt the "principle of mediocrity" by assuming planets to be a usual byproduct of star formation. They develop powerful arguments that reasonably show how solar systems and planets might develop from interstellar gas and dust.

This cosmic scenario has been developed in greater detail by others, but the essentials remain. A vast cloud of interstellar gas and dust gravitationally contracts, increasing its majestic galactic spin. It fragments into smaller and smaller collapsing clouds until protostar systems ranging in size from 0.1 M_O to 1000 M_O evolve. While the central protostar contracts still further, accretes more mass, and becomes hotter, the planets evolve by accretion of the small residue of dust and gas that was not gathered by the protostar. Ices, frozen gases, organic molecules, silicates, irons, and other minerals become glued together by gravitational, magnetic, electrostatic, and organic adhesive forces. Ever larger chunks of matter orbit the protosun, each interfering with the other and further accreting by chance contacts until protoplanet-size bodies are formed. These larger bodies continue accretion by collision—witness the extensive cratering seen on Mercury, the Moon, Mars, and even (as Raymond Siever shows) on the weathered face of the Earth. Finally, some planets (such as Venus and the Earth) become so large that the heat generated by radioisotope decay builds up faster than it can be dissipated from the surface. The interior melts and becomes fractionated: denser materials such as iron and nickel sink to form a molten core, while lighter materials float to the surface to form the mantle and crust. Smaller bodies such as Mercury, the Moon, and Mars seem never to have experienced high enough interior temperatures to form large well-differentiated cores.

Thus we have a picture of planets formed by accretional processes. They become internally fractionated and melted only if large enough. As a result of the process of internal heating, volcanic emissions release volatile materials that become the first atmosphere.

At some early stage of planet formation, a brief, explosive increase in protosun luminosity takes place. This drives away material equivalent to some two to three times the solar system's present mass. The relationship of this T-Tauri expulsive stage, which we can observe directly in nebular clouds containing protostars, to any specific stage of planet formation is uncertain. If the T-Tauri phenomenon occurred before planetoid aggregation, then most of the original frozen volatiles (carbon dioxide, water, nitrogen, and so on) frozen onto nebular dust would have been lost, and no future atmosphere could have been formed by volcanic outgassing. If the T-Tauri stage occurred after planet formation, however, the original volcanically-produced atmosphere would have been lost. But the Earth and Venus have atmospheres that can be traced back to volcanoes, and Mars and the Moon, which have at best tenuous or nonexistent atmospheres, have had little internal heating. Therefore we might reason that the T-Tauri loss of volatiles occurred after primary planetoids formed, but before final planet creation. This is speculation; considerable geological and astronomical research remains to be done to elucidate this point. Some researchers have even speculated that *all* volatile materials were lost before planetoid formation. This hypothesis requires later addition of volatiles by some other mechanism, further confusing the issue. Were this hypothesis true, planetary atmospheres might be dependent upon chance collisions of the protosystem with comets or volatile-rich nebular clouds. This would make the formation of the Earth's early atmosphere a rare, chance event. The probability of similar events for other protosystems is low, and we—life—become a unique happening. At this time the reader can choose a mental model of planet formation on the basis of insufficient, but fascinating, data. Good luck.

Suggested Further Reading

Jastrow, Robert, and A. G. W. Cameron (eds.). 1963. *Origin of the Solar System.* Academic Press, New York.

Kummel, Bernhard. 1970. *History of the Earth: An Introduction to Historical Geology,* second edition. W. H. Freeman and Company, San Francisco.

Press, Frank, and Raymond Siever. 1978. *Earth,* second edition. W. H. Freeman and Company, San Francisco.

Scientific American Book, A. 1975. *The Solar System.* W. H. Freeman and Company, San Francisco.

Urey, Harold C. 1952. *The Planets, Their Origin and Development.* Yale University Press, New Haven.

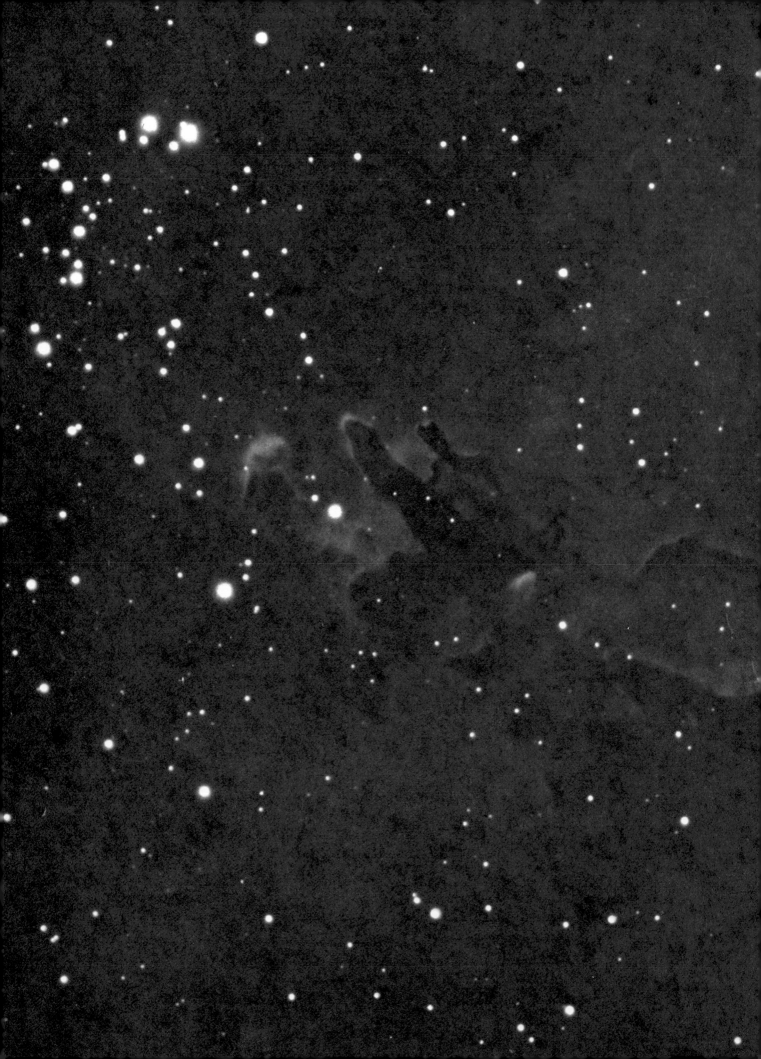

1

The Origin and Evolution of the Solar System

by A. G. W. Cameron
September 1975

It is generally agreed that some 4.6 billion years ago the sun and the planets formed out of a rotating disk of gas and dust. Exactly how they did so remains a lively topic of investigation

A great cloud of gas and dust contracted through interstellar space 4.6 billion years ago, far out along one of the curved arms of our spiral galaxy. The cloud collapsed and spun more rapidly, forming a disk. At some stage a body collected at the center of the disk that was so massive, dense and hot that its nuclear fuel ignited and it became a star: the sun. At some stage the surrounding dust particles accreted to form planets bound in orbit around the sun and satellites bound in orbit around some of the planets.

So goes—in very broad outline—the nebular hypothesis of the origin of the solar system. Its central idea was proposed more than 300 years ago. It sounds simple enough, and it makes intuitive sense to the layman; indeed, some version of it is accepted by most astronomers today. And yet beyond the broad outlines there is no consensus among students of the origin and evolution of the solar system. We still have no generally accepted theory to explain how the primitive solar nebula formed, how and when the sun began to shine and how and when the planets coalesced out of swirling dust.

It was René Descartes who first proposed (in 1644) the concept of a primitive solar nebula: a rotating disk of gas and dust out of which the planets and their satellites are made. A century later (in 1745) Georges Louis Leclerc de Buffon put forward a second theory: that a massive body (he suggested a comet) came close to the sun and ripped out of it the material that constituted the planets and their satellites. In the two centuries after Buffon the many theories that were propounded tended to follow in the tradition of either Descartes's monistic view or Buffon's dualistic one; the balance of favor swung back and forth between them. The most significant early monistic theories were those of Immanuel Kant and Pierre Simon de Laplace, who elaborated on Descartes's original idea by explaining how the cloud of gas and dust, shrinking to form the sun, would have spun faster and faster because of the conservation of angular momentum: a decrease in the radius of a rotating mass must be balanced by an increase in its rotational speed. Laplace suggested that a series of rings were shed, from whose dust the planets and satellites were formed. At the end of the 19th century dissatisfaction with the ability of the nebular hypothesis to explain the accretion of matter into the planets brought dualistic theories back into favor. Today they have been generally abandoned; it seems clear that most of the material that might have been drawn out of the sun by, say, the approach of another star would have fallen back into the sun or dispersed in space before any solid condensates could coagulate into planets.

A major reason for the wide range of early theories of the origin of the solar system was the lack of observational data—of facts to be explained by a theory. The history of the earth's first few hundreds of millions of years is missing from the geological record, which could therefore offer no clues to the environment in which this sample of a planet was born, and the limited capabilities of telescopes restricted the astronomical data. The early theories were devised to explain only a few observations: the spacing of the planetary orbits increased in a regular way (in accordance with what is known as Bode's law); planetary orbital motions and spins tended to have the same direction of rotation; the sun accounted for only a small fraction of the total angular momentum of the solar system, even though it accounted for the greatest fraction by far of the total mass of the system. These few facts provided few constraints on theory, and so the theories proliferated.

In just the past three decades the situation has changed dramatically. We have a vast amount of new information that imposes additional and powerful constraints on any theory. The new knowledge stems notably from new research on meteorites and from the data returned to the earth by spacecraft dispatched to other bodies in the solar system.

The meteorites are samples of primitive solar-system material. They are evi-

NEW STARS ARE BORN, as the sun may have been born, in gaseous emission nebulas: diffuse, dusty clouds of hot interstellar gas. The photograph on the opposite page, made with the Mayall four-meter reflecting telescope at the Kitt Peak National Observatory in Arizona, shows the nebula designated M16 or NGC 6611, called the Eagle Nebula, in the local arm of our galaxy. Within the nebula clouds of gas have condensed relatively recently to form bright blue-white stars, and other such clouds are still condensing. Ultraviolet radiation from the hot new stars ionizes hydrogen atoms in the remaining gas, giving rise to free electrons and protons. When high-energy electrons recombine with protons, light is emitted at the hydrogen-alpha wavelength of the spectrum: red light that illuminates the cloud and silhouettes dense, dusty, cooler regions of the nebula where the light does not penetrate.

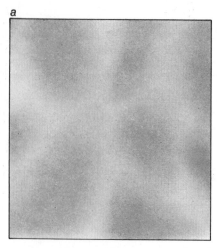

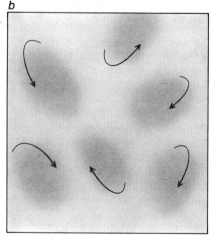

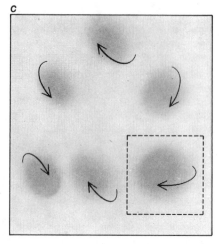

GALAXIES FORMED in the thin, expanding primordial gas (mostly hydrogen, with some helium) when regions of somewhat greater density (*a*) contracted gravitationally to form protogal- axies (*b*), rotating because of the net effect of gas eddies within them. The protogalaxies continued to contract gravitationally, and then to rotate faster (*c*). One of them (*rectangle*) was our own.

dently fragments of rather small bodies that have collided and broken up, sending many of their pieces into new orbits that ultimately intersect the earth. They bring to us, trapped in their interior, samples of the gases of the solar nebula. The details of their mineralogy provide clues to the temperatures and pressures in the nebula at the time its individual grains were last exposed to chemical reaction with its gases. From the relative amounts of the products of radioactive decay that remain trapped in the interior of the meteorites we learn how long ago the original elements that gave rise to certain radioactive isotopes were assembled to form the meteorites' parent bodies.

One of the primary scientific goals of the space-probe program was to advance understanding of the origin of the solar system, and the program has already borne fruit. Measurements made by spacecraft have refined our knowledge of planetary masses and radii, from which we derive accurate mean densities of the planets and clues to their internal composition. By observing how the gravitational potentials of a planet differ from those of a perfectly uniform sphere we derive constraints on the degree to which the density can vary in different parts of the planet's interior. Determining whether or not a planet has an intrinsic magnetic field tells us something about the planet's internal dynamics. Spacecraft data on the composition of a planetary atmosphere reveal something about the gases that once were incorporated in the planet and about chemical interactions between the atmosphere and the planet's surface. Examining the incredibly detailed images of solid plane-

tary surfaces that have been sent back by spacecraft cameras, we can see how volcanic and other geological processes have operated on other planets. The density of craters tells us about the terminal stages of the planet's accretion and about the numbers of smaller bodies that have wandered through the solar system.

Still other constraints come from the general advances in astrophysics that have marked the past three decades. We now know that our galaxy as a whole is between two and three times older than the solar system; we therefore have good reason to believe that the conditions we see today in the galaxy are not very different from those at the time the solar system was formed. We see regions in our galaxy in which stars have been formed in the recent past and are probably still being formed today; that gives us important information if we believe the sun and the solar nebula formed as parts of the same general process. We have learned much about the birth and death of stars and how elements originate in nuclear reactions within exploding stars and are formed into tiny grains of interstellar dust, and about how those grains concentrate in the dark patches in the sky that blot out the light coming to us from distant stars. Those grains of dust and the interstellar gases that accompany them were the raw material of the solar nebula. Let me now try to weave the many threads of information into a coherent picture of the solar system's formation.

Galaxies form when gas—mostly hydrogen—collapses out of intergalactic space. Many billions of years before

the origin of the solar system our galaxy began to take shape in that way. Out of the collapsing gas a first generation of stars was born—stars that still remain spherically distributed around the center of the galaxy, a reminder of its original roughly spherical shape. After those first stars were formed the residual gas, because of its intrinsic angular momentum, settled into the thin disk that is a characteristic feature of all spiral galaxies, and further generations of stars formed from the gas in the disk. The more massive of them evolved quickly, forming heavy elements that were ejected into the interstellar gas. Some of the heavy elements condensed into tiny grains: the interstellar dust. When enough stars had formed in the central plane of the galaxy, an instability de-

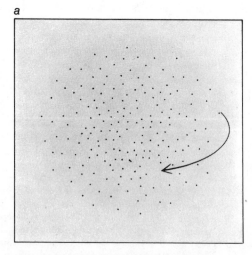

OUR GALAXY EVOLVED as dense lumps of gas contracted within the protogalaxy to form a first generation of stars (*a*). In time the residual gas settled into a disk in the

veloped in their motions that allowed them to cluster together temporarily, forming the spiral arms.

Such arms represent local enhancements of star-population density in the disk; the arms are continuing features that rotate around the center of the galaxy, but the material that constitutes them keeps changing: individual stars spend only about half of their time in one arm before moving on to the next one. Like the stars, the interstellar gas and dust spend about as much time in an arm as they do flowing through the larger spaces between successive arms; the result is that the density of gas and dust is considerably enhanced in a spiral arm. We know from studying galaxies other than our own that it is in these high-density spiral arms that the new stars of spiral galaxies are formed.

Pressure differences arise within the gas, perhaps as the result of a supernova explosion; the gas flows away from regions of higher pressure, but in moving it may tend to pile up somewhere else. Clouds of high-density un-ionized gas accumulate, which typically have a mass of from several hundred to several thousand times the mass of the sun. Gravitational forces tend to pull such a cloud into a more compact configuration. Contraction is opposed, however, by the internal pressure of the gas in the cloud, which tends to make the cloud expand; ordinarily the internal pressure is much stronger than the gravitation and the cloud is in no danger of collapsing. Sometimes, however, a sudden fluctuation in pressure—from a nearby violent event such as a supernova explosion, the formation of a massive star or a large re-

arrangement of the interstellar magnetic field—may compress a cloud to a density much higher than normal. Under such conditions, which are quite rare, gravity may win out over internal pressure, so that the cloud begins to collapse to form stars. As the cloud collapses, its interstellar grains shield its interior against the heating effect of radiation from the stars outside. The temperature of the cloud falls, and the internal pressure becomes less effective. The collapsing cloud breaks into fragments and the fragments break into smaller fragments. When one small fragment eventually completes its collapse, it will have formed into a flattened disk, cool at the edges and very hot at the center: a primitive solar nebula.

What was the nature of the solar nebula and how did it evolve? When did the sun form? Why are there planets? How did they take shape? Quite different pictures of the structure of the primitive solar nebula and of its evolution result from different estimates of its size. Such estimates have usually been arrived at by reasoning backward in time from the present masses of the planets. Let me reproduce such an argument.

In a very general way one can divide the materials of the planets into three classes depending on their volatility: rocky, icy and gaseous. The major constituents of rocky materials are iron and oxides and silicates of magnesium and other metals, notably aluminum and calcium. All these materials would be in solid form at pressures characteristic of the primitive solar nebula and at tem-

peratures in the range from 1,000 to 1,800 degrees Kelvin (degrees Celsius above absolute zero). The four inner planets and the earth's moon (and at least two of the major satellites of Jupiter) appear to be basically rocky. The rocky solids represent about .44 percent by mass of the material out of which the sun formed. The present mass of a rocky planet, then, represents about .44 percent of its share of the primitive solar nebula; the remainder of that share is "missing" because it was too volatile to have been incorporated in the planet.

At a temperature below 160 degrees K. the water in the nebula would be in the form of ice. Ammonia and methane form solids only at a somewhat lower temperature. The ices constitute 1.4 percent by mass of the material out of which the sun formed. Rock-ice mixtures account for most of the mass of Uranus and Neptune and some of the mass of Saturn and Jupiter (and for the bulk of the mass of most of the satellites of the outer planets and the comets). Arguing as for the inner planets, one can assume that the rock and ice now present in such bodies represent 1.4 plus .44 percent, or 1.84 percent, of those bodies' original share of the nebula.

At any temperature likely to have been attained in the primitive solar nebula the very volatile elements—hydrogen and the noble gases such as helium and neon—would remain in the gaseous state. Such gases are incorporated in bodies within the solar system only to the extent that they have been held in planetary atmospheres by gravity and, in the case of hydrogen, held in chemical compounds such as water. (A tiny amount

b

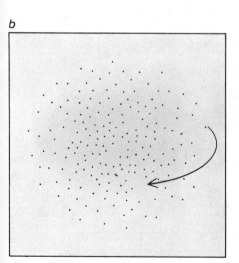

c

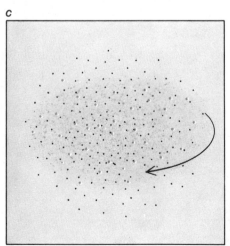

d

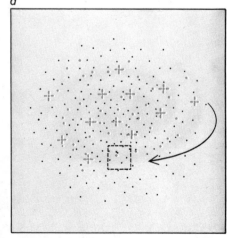

plane of the galaxy's rotation under the combined influence of gravity and centrifugal force (b). Further generations of stars (color) formed within the disk (c); some stars, evolving rapidly, produced heavier elements through nuclear fusion and ejected them into the disk, where some elements condensed into solid interstellar grains. Instabilities in the motions of gas and stars led to density enhancements that we see as the spiral arms of the galaxy (d). The area in the rectangle is enlarged in the first drawing on the next page.

of helium comes from the decay of radioactive elements.) Morris Podolak and I recently analyzed the structure of the outer planets. We determined that hydrogen and helium constitute about 15 percent of Uranus' mass, about 25 percent of Neptune's, about two-thirds of Saturn's and about four-fifths of Jupiter's. In these planets it is necessary to allow for the gaseous components in order to establish the present rock-ice mass.

By thus establishing the rock and the rock-ice masses of the planets and augmenting those masses for the missing constituents that were too volatile to condense it is possible to estimate a minimum mass for the primitive solar nebula: a mass sufficient to account for the formation of the planets. That minimum mass is about 3 percent of the mass of the sun [see illustration on opposite page]. (Older estimates arrived at a much smaller mass—less than 1 percent

of the sun's—because they did not allow for enough rock and ice in Jupiter and Saturn.)

The 3 percent figure is definitely a minimum. It assumes that the planets were completely efficient in collecting from the solar nebula all the material that was in condensed form in each planet's orbit in the nebula. For two kinds of solid, however, that collection process might have been quite inefficient. Consider first the tiny unconsolidated grains of interstellar dust, perhaps a micrometer (a thousandth of a millimeter) in diameter, that were not vaporized as the gas-cloud fragment collapsed. The thickness of the nebular disk must have been at least one astronomical unit (the mean distance between the earth and the sun). That dimension is very large compared with the dimensions of any of the planets, which consolidated approximately in the central plane of the disk. Gas-drag effects

would prevent large quantities of these small grains from settling through the nebular gas toward the central plane at a significant rate; if much of the gas was instead dissipated inward to form the sun, the grains would have accompanied the gas and could never have become incorporated in the planets.

Larger bodies (centimeters or meters in diameter), on the other hand, would fall rapidly through the gas toward the midplane but might nevertheless not end up in planets. As a result of a difference between the centrifugal forces that act on the solid bodies and those that act on the gas, the solids would rotate around the central spin axis of the nebula more rapidly than the accompanying gas. They would therefore move through the gas with a relative velocity as high as several hundred miles an hour; a head wind of that speed would tend to slow them down so that they would spiral rather quickly through the gas toward

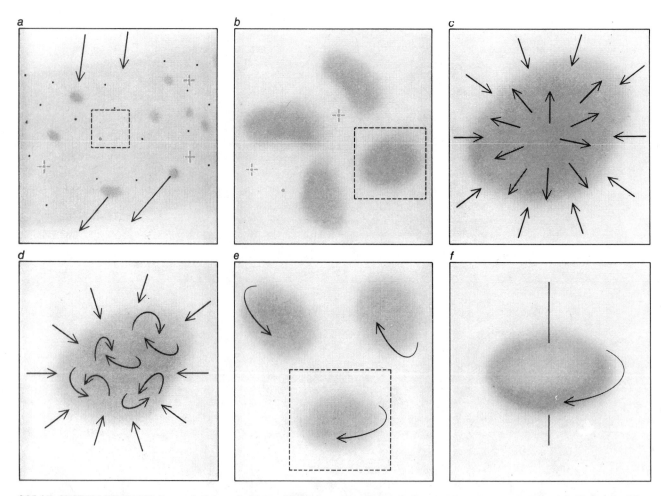

SOLAR SYSTEM EVOLVED in a spiral arm about two-thirds of the way out from the center of the galaxy. Stars, gas and dust grains move through the arm, and new stars are born there; massive, short-lived stars outline the arms (a). A supernova explosion or the birth of massive stars creates instabilities that concentrate high-density clouds of gas (b). Gravitational forces contract the cloud, but the cloud's internal pressure opposes contraction (c). If the cloud has enough mass, gravity dominates (d) and the cloud collapses. The collapse generates strong gas eddies (curved arrows) and breaks the cloud into fragments (e); each fragment has a net rotation derived from its major eddies. One of these fragments spins faster and its gas settles into a disk that was the primitive solar nebula (f).

the central spin axis and thus be lost to the region of planet formation. For these two reasons the mass of the primitive solar nebula may have been considerably larger than 3 percent of the sun's mass.

Many of the solar-system theories constructed over the past three decades have involved some version of a minimum-mass solar nebula. The concept has a major flaw, however. It assumes that the sun itself was formed directly during the process of collapse and that the primitive solar nebula was marshaled independently around the sun. The trouble is that simple estimates of the amount of angular momentum that must have been contained in the collapsing cloud fragment indicate that it would have been impossible for almost all of the fragment simply to collapse directly to form the sun, leaving a small fringe of nebula to constitute the planets. Such estimates require instead that the nebula's mass be spread out over several tens of astronomical units. The solar nebula itself must have contained substantially more than one solar mass—and probably about two solar masses—of material, with no sun originally present at the central spin axis. Let me first explain the source of the large amount of angular momentum and then show why it indicates that there was not a minimum solar nebula but a massive one.

The strong fluctuations in pressure that led to the rapid compression of the original interstellar cloud and thus brought it to the threshold of gravitational collapse must have stirred the cloud's gases into violent turbulence. Large-scale shearing motions developed —eddies superimposed on eddies, in a wide range of sizes and in many planes and directions. When any one fragment became isolated from such a turbulent cloud, it had a net tendency to spin, derived from the motions of the largest eddies it happened to contain. A fragment's mass, its rate of rotation and its radius combine to endow it with a certain amount of angular momentum, and that momentum must be conserved; as the fragment contracted, it spun faster. The sun turns very slowly, however; in spite of its great mass it accounts for only 2 percent of the solar system's angular momentum. Most of the original angular momentum of the vast quantities of gas that moved in to form the sun must have been transported outward; a considerable part of the original nebula must therefore have remained at

PLANET	PRESENT MASS (PERCENT OF SUN'S)	AUGMENTED MASS (PERCENT OF SUN'S)
MERCURY	.000017	.004
VENUS	.000245	.056
EARTH	.000304	.07
MARS	.000032	.007
JUPITER	.09547	1.5
SATURN	.02859	.77
URANUS	.00436	.27
NEPTUNE	.00524	.27
PLUTO	.00025 (?)	.06 (?)
TOTAL (MINIMUM MASS OF SOLAR NEBULA)		3.0

MINIMUM MASS OF SOLAR NEBULA is estimated by adding up the amount of solar material that must have been present (*column at right*) to account for the present mass (*middle column*) of each planet. Solar-nebula mass thus estimated is 3 percent of mass of sun.

great distances from the sun to take up that angular momentum.

An additional reason for postulating a massive solar nebula is the observation that young stars tend to lose mass at a prodigious rate early in their lifetime; the loss comes as they pass through what is called their T Tauri stage, which I shall discuss in a bit more detail below. The combination of the mass that remained in the solar nebula and never became part of the sun and the mass that was once in the sun but was lost in the early sun's T Tauri stage could easily have amounted to as much as one solar mass.

As a result of this kind of reasoning—in effect arguing forward from what is known of the principles of star formation rather than backward from the masses of the present planets—Milton R. Pine and I constructed some numerical models of the massive solar nebula. The models extended out to a radial distance of about 100 astronomical units and contained two solar masses of material. In a typical model the temperature was about 3,000 degrees K. near the spin axis and decreased to a few hundred degrees in the region of planet formation. Such temperatures are considerably higher than the temperatures that characterized the collapse of the original interstellar cloud; they develop in the later stages of compression of the gas, once its density becomes high enough so that its own cooling radiation can no longer escape easily. The escape of this radiation is impeded, however, only during the rapid final stages of the collapse; once the gas stops contracting— once the primitive solar nebula is formed

—the radiation can escape relatively quickly, so that in the region of planet formation the nebula will lose most of its heat energy in only a few hundred or a few thousand years.

Such a short cooling time (short compared with the time required to form a sun and planets) presents a difficulty for the massive-nebula model. As the nebula cools it will flatten into a thinner disk, and thin disks have been shown to be dynamically unstable: they tend to deform into a barlike configuration. (Such a deformation might well be the mechanism by which close pairs of double stars are formed, but that evidently did not happen in the solar system.)

There is another time-scale problem for the massive-nebula model. An important process for transporting angular momentum away from the central spin axis so that gas can shrink toward that axis is probably a system of fast meridional currents: gas currents that flow in a plane parallel to the spin axis and at a right angle to the central plane of the nebula. Pine and I estimated that the characteristic time for the outward transport of the angular momentum shed by the inner parts of the primitive solar nebula would be only a few thousand years. John Stewart of the Max Planck Institute for Physics and Astrophysics in Munich has shown that gas turbulence must play an important role in a primitive solar nebula and may cause an even more rapid outward transport of angular momentum.

Both the time for cooling and the time for angular-momentum transport seem too short compared with the time required for the accretion of the solar nebula. After a fragment separates from the interstellar cloud its central region is likely to be denser, and will collapse more rapidly, than the remainder of the fragment. A small solar nebula will therefore be formed at first when the central region ceases to collapse; that small nebula will grow by accretion of the remainder of the infalling fragment

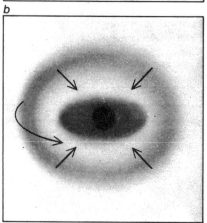

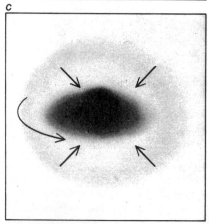

ACCRETION MODEL of the primitive solar nebula assumes that a central region of the cloud fragment collapses faster than the rest (a). It forms a small solar nebula: a central mass that is not yet the sun, surrounded by a disk of gas and dust grains, with more gas and dust concentrated around the periphery (b). The small nebula then grows by accretion over a long period of time (c).

over a period of time—probably between 10,000 and 100,000 years, a lot longer than the cooling and angular-momentum transport times we estimated.

These considerations have led me to a new picture of the primitive solar nebula that I am currently trying to define in detail. It is necessary to construct not a single model but an evolutionary sequence of models, beginning with a small solar nebula that grows through accretion over a time interval of perhaps 30,000 years. In this case the time for the redistribution of angular momentum remains short compared with the accretion time, so that much of the mass flows inward to form the sun not at the beginning of the accretion period but throughout that period; the mass of the nebula out in the region of planet formation remains a relatively small fraction of a solar mass throughout the period. As for cooling, the accreting gas is suddenly decelerated when it hits the surface of the solar nebula; the energy of its infall is converted into heat that is radiated from the surface. In the later stages of accretion that process keeps the surface layers in the region of planet formation at a temperature of perhaps a few hundred degrees; the temperature in the interior would be somewhat higher. And meanwhile the steady flow of mass toward the central spin axis diminishes dynamic instabilities within the nebula.

The two pictures of the primitive solar nebula, one derived from the masses of the planets and the other from the principles of star formation, thus seem to be converging to form an intermediate model of the initial solar nebula. In that model somewhat more than one solar mass has collected toward the spin axis but is not yet recognizable as the sun. It is surrounded by a disk of gas and dust amounting to perhaps a tenth of a solar mass. Farther out, beyond the region of planet formation, considerable additional amounts of mass are still falling toward the solar nebula.

The planets were created by the accumulation of interstellar grains and, in the case of the outer planets, the subsequent attraction and adherence of gases. The buildup of solid matter would have begun, I have recently calculated, in the collapsing gas cloud. Turbulent gas eddies would have accelerated the interstellar grains until they had large enough relative motions to begin to collide with one another. Having been formed out of material in stars and then ejected into interstellar space, where ices and other volatile constituents condensed on their

surface, the grains probably had a rather fluffy structure. It would not be surprising if such particles stuck to one another when they collided, forming clumps. As time passed the clumps of grains would collide with one another, sometimes amalgamating into larger clumps and sometimes breaking up into smaller ones. By the time the solar nebula had formed, many clumps were likely to have grown to a diameter measured in millimeters or centimeters.

Clumps of that size could settle through the gas toward the midplane of the nebula in tens or hundreds of years. Since their settling rate would vary with size there would be further collisions, increasing the size of the clumps and accelerating their fall toward the midplane. At that point, however, unless they were somehow able to grow substantially larger they would rapidly be lost to the inner solar nebula as a result of the gas-drag effect I mentioned above.

A critical process in planet formation may therefore be a mechanism recently proposed by Peter Goldreich of the California Institute of Technology and William R. Ward of Harvard University, which would give rise to those larger bodies. They showed that if there is a thin layer of condensed solids at the midplane of the nebula, with very little relative velocity among the particles, then a powerful gravitational instability mechanism will break up the thin sheet into bodies with diameters in the range of the diameters of asteroids: kilometers or tens of kilometers. The instability mechanism gradually operates over larger distances, attracting the asteroid-size bodies into loosely bound gravitating clusters of hundreds or thousands of bodies. The clusters remain unconsolidated because of the large angular momentum contained in their component bodies, which makes them rotate around common gravitating centers. When two clusters approach each other, however, they intermingle; the fluctuating gravitational field in the combined cluster leads to a violent dynamic relaxation of the motions of the bodies, so that many of them coalesce to form cores around which others go into orbit (although some of the bodies would be lost). The clusters interact with one another gravitationally over quite large distances; mutual perturbations gradually build up the velocities of the clusters with respect to one another, leading to further collisions that produce ever larger bodies.

The Goldreich-Ward instability mechanism would appear to be a powerful first step in the accumulation of planetary bodies. The subsequent stages in

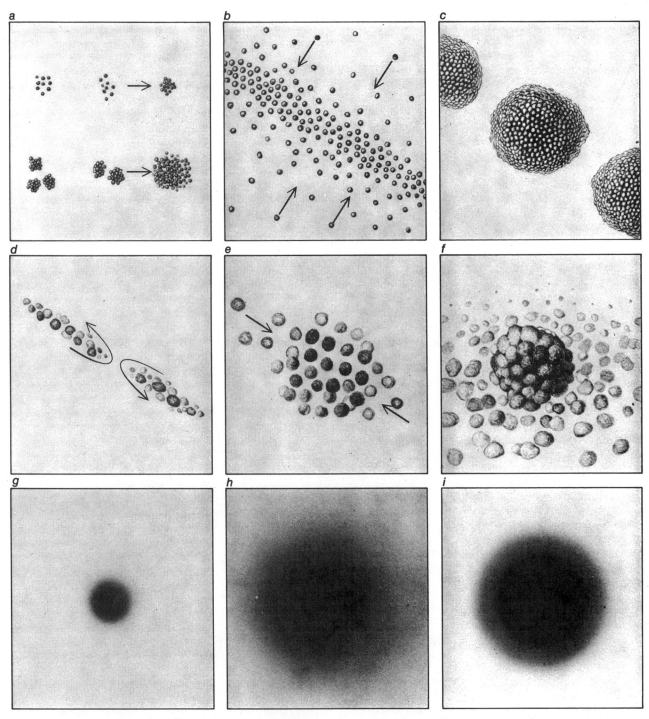

PLANETS BEGIN TO FORM when interstellar dust grains collide and stick to one another, forming ever larger clumps (a). The clumps fall toward the midplane of the nebula (b) and form a diffuse disk there. Gravitational instabilities collect this material into millions of bodies of asteroid size (c), which collect into gravitating clusters (d). When clusters collide and intermingle (e), their gravitational fields relax, and they coagulate into solid cores, perhaps with some bodies going into orbit around the cores (f). Continued accretion and consolidation may create a planet-size body (g). If the core gets larger, it may concentrate gas from the nebula gravitationally (h). A large enough core may make the gas collapse into a dense shell that constitutes most of the planet's mass (i).

the process are still highly speculative and were surely different in different regions of the solar nebula. The interstellar grains whose clumping initiates the accumulation process are those that have not been vaporized by the heat of the nebula; their materials, and therefore the materials of the larger bodies

into which they are incorporated, would be different at different distances along the steep temperature gradient: metals, oxides and silicates in the region of the inner planets; similar rocky compounds and water ice farther out; rock, water ice and frozen methane and ammonia still farther out.

In the case of the smaller inner planets the progression to full size may be just a question of successive collisions and amalgamations of rocky bodies. In the case of the outer planets there are other considerations. Fausto Perri and I have recently considered the behavior of the primitive solar nebula as a large planetary core grows within it. As the mass of the core increases, gas in the solar nebula becomes gravitationally concentrated toward the core; with the continued growth of the core the amount of mass in the gas that is concentrated increases even more rapidly than the mass of the core itself. At some point the core reaches a critical size (which depends on the temperature conditions in the surrounding gas) such that the gas becomes hydrodynamically unstable and collapses onto the planetary core.

The major constituents of Jupiter and Saturn are the hydrogen and helium in their atmosphere, and we believe it was through this process of concentration and collapse that these planets acquired most of their mass. Hydrogen and helium account for a smaller fraction of the mass of Uranus and Neptune, probably indicating that their core never grew to the critical size for hydrodynamic collapse; those two planets did, however, grow large enough to retain much of the hydrogen and helium that was gravitationally concentrated toward their core. The inner planets, on the other hand, may be too small ever to have concentrated much of the nebular gas.

When the collapse events took place to form Jupiter and Saturn, local conservation of angular momentum in the gas would cause it to flatten into a disk around the planetary core. As time went on the two planets would sweep up essentially all the gas in their vicinity within the nebula. One can think of them as forming miniature versions of the primitive solar nebula: a central core of condensed rock and ice taking the place of the sun, with the gaseous disk around the core as the analogue of the

solar nebula. Both of these large planets have systems of regular satellites incorporating considerable mass, which probably formed from a gaseous disk by processes quite analogous to the formation of the planets in the solar nebula.

Any theory of the origin and evolution of the solar system must account somehow for the comets, its most spectacular but least understood members. Jan Oort of the Leiden Observatory suggested some years ago that the comets inhabit an enormous volume of space centered on the sun, starting well beyond the outer planets and extending to a distance of perhaps 100,000 astronomical units. The total mass of the comets in this vast "Oort cloud" is probably equivalent to between one earth mass and 1,000 earth masses, which would account for between 10^{12} and 10^{15} comets. The comets we see are those few whose orbital elements are perturbed by a passing star in just the right way to send them plunging toward the center of the solar system.

A comet is a "dirty snowball," an aggregate of ice and rocky material, in the model first suggested by Fred L. Whipple of Harvard University. As a comet approaches the sun, gases are vaporized from it, accompanied by dust particles, to form the characteristic coma and tail. Analysis of the tails shows that the molecules that are vaporized are primarily water but also include exotic organic compounds. The dust, some of which comes to rest high in the earth's atmosphere, consists of fluffy clumps of fine-grained rocky material. A comet, in other words, is apparently an assembly of interstellar grains.

The comets must either have been made within the solar nebula and somehow ejected into the Oort cloud or else have been made out in the Oort cloud itself. Oort originally suggested that they were formed near Jupiter and perturbed by Jupiter's gravitational field into very large orbits that were subsequently rounded out by stellar perturbations. That would require the formation of a staggeringly large mass of comets, since many more would have been ejected from the solar system than were retained in the Oort cloud. Moreover, given the temperatures that must have prevailed near Jupiter it seems unlikely that molecules more complex than water would have been in solid form. More complex ices are possible farther out in the nebula, and so Whipple and others have suggested that the comets were formed in the neighborhood of Uranus and Neptune and sent out into the Oort cloud by the gravitational fields of those planets. That proposal, however, meets only one of the objections to the Oort hypothesis.

My own belief is that the comets were probably formed out in the Oort cloud itself. It is true that the collapsing gas of the fragment of cloud that became the primitive solar nebula was never dense enough so far from the center for the interstellar grains to have aggregated into sizable bodies out there. There is another possibility, however. Most of the stars in the galaxy are much less massive than the sun, suggesting that gas-cloud fragmentation sometimes continues at least down to fragments a tenth of a solar mass in size. Fragmentation may have gone further still. Small fragments could have been bound gravitationally to the primitive solar nebula in

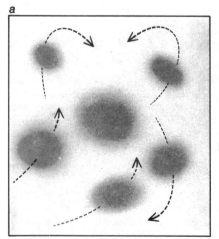
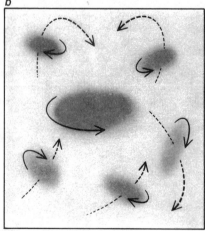
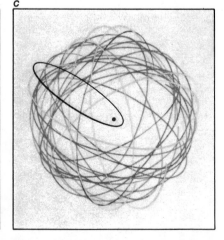

COMETS MAY HAVE FORMED from small cloud fragments that once were in orbit around the larger fragment that became the solar nebula (a). The small fragments spun down, like the solar one, to form disks in which comets were accumulated much as planets were (b). Eventually starlight could have evaporated the gases of these "cometary nebulas," leaving the comets in enormous orbits around the sun (c). From time to time a comet's orbit is perturbed by a passing star, and the new orbit brings it close to the sun.

a

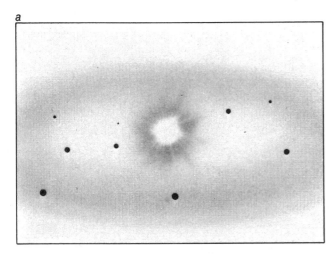

b

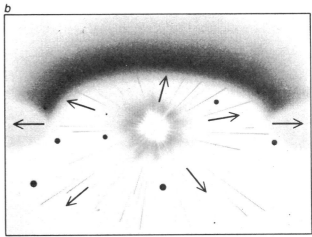

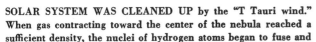

SOLAR SYSTEM WAS CLEANED UP by the "T Tauri wind." When gas contracting toward the center of the nebula reached a sufficient density, the nuclei of hydrogen atoms began to fuse and the sun began to shine (*a*). During its T Tauri phase the sun lost vast quantities of material. That material constituted an intense solar wind that could have blown away the remaining gas (*b*).

orbits traversing the region of the Oort cloud. Such fragments would form fairly large and very cool disks, ideal places for the comets to form. When the disks were ultimately heated by ultraviolet radiation from external stars, the gases would evaporate away and leave the comets in solar orbits [*see illustration on opposite page*].

After the planets had formed, much of the gas of the solar nebula must have remained in orbit around the sun, along with countless small bodies and large amounts of unconsolidated dust. There are only planets and asteroids in orbit now, with very little dust and almost no gas. How was the solar system cleaned up? As I mentioned above, young stars characteristically pass through what is called the T Tauri stage, when they eject matter at a prodigious rate: as much as one solar mass per million years! There is every reason to believe the sun passed through a similar phase, and the fierce "wind" of that ejected mass undoubtedly dissipated the solar nebula by carrying the residual gas off into space. That early solar wind would have stripped the inner planets of the remains of any primitive atmosphere of hydrogen and helium from the primitive solar nebula; the outer planets must have formed early enough to capture their hydrogen and helium before the solar wind began to blow. Moreover, if the accretion of gas from the original cloud fragment was still continuing, it would have been terminated by the wind, with the infalling gases being ejected back into interstellar space.

What determined when the T Tauri wind began? Why did it not blow away the primitive solar nebula long before the sun got so large? The thermonuclear reactions of hydrogen that constitute a stellar furnace are ignited only under extreme conditions of high temperature and high density. Perri and I have recently determined that the temperatures of the primitive solar nebula were so low that compressional heating of the gas could not have ignited the sun at a central density comparable to that of the sun today; the density would have had to be at least 100 times higher than it is now. Only then could the sun have adjusted itself into its present configuration, and only then could the thermonuclear furnace have been ignited and could the intense T Tauri wind begin to blow. An enormous amount of mass must have had to be gathered together to achieve such a density. Given the original temperature of the nebula, in other words, the sun had to reach a large mass before it could be a sun.

Once the sun had begun to shine and the T Tauri wind had blown away the gas, the stage was set for the final cleanup of interplanetary space and the completion of planet formation. The orbits of most of the small bodies in the solar system (other than the observed asteroids in their isolated belt) would have been continually modified by planetary perturbations. Over the course of a few hundred million years most such bodies either would have collided with one of the planets (the surfaces of Mercury, the moon and Mars still show the scars of that terminal bombardment) or would have been ejected from the solar system by a major planet, usually Jupiter.

The tiny dust particles were subjected to forces even stronger than gravitational perturbations: the effects of sunlight. The photons the particles absorb from the sun carry no angular momentum; the photons the particles radiate, however, carry off some of the angular momentum of the particles' orbital motion. The sunlight therefore acts as a resisting medium for the particles, making them spiral in toward the sun. Larger solids, up to a kilometer in diameter, are perturbed by sunlight in a different way. As such a body rotates the temperature of a section of its surface increases as long as it is on the sunlit side but decreases while it is on the dark side. One hemisphere of the body therefore emits considerably more radiation than the other. That gives rise to a preferential thrust that can perturb the orbit of the body either toward the sun or away from it, depending on the body's direction of rotation. Such bodies will eventually come close to one of the planets, whereupon they will be absorbed by collision or be ejected from the solar system. In these several ways the sunlight could have acted as a broom to sweep away much of the smaller debris left over from the formation of the solar system.

The account I have given makes a coherent, if incomplete, story. Many of its details remain highly speculative, however, and much of the story may have to be retold as new data test the present theories. In selecting and weaving together facts, ideas and hypotheses I have necessarily been strongly influenced by my own beliefs. The reader should be aware that others would weave quite different tapestries. These ancient questions are still far from being answered.

2

The Origin
of the Earth

by Harold C. Urey
October 1952

*The emergence of the theory that the solar system
coagulated from a vast cloud of dust has led to a new
inquiry into the chemical history of our planet*

IT IS PROBABLE that as soon as man acquired a large brain and the mind that goes with it he began to speculate on how far the earth extended, on what held it up, on the nature of the sun and moon and stars, and on the origin of all these things. He embodied his speculations in religious writings, of which the first chapter of *Genesis* is a poetic and beautiful example. For centuries these writings have been part of our culture, so that many of us do not realize that some of the ancient peoples had very definite ideas about the earth and the solar system which are quite acceptable today.

Aristarchus of the Aegean island of Samos first suggested that the earth and the other planets moved about the sun—an idea that was rejected by astronomers until Copernicus proposed it again 2,000 years later. The Greeks knew the shape and the approximate size of the earth, and the cause of eclipses of the sun. After Copernicus the Danish astronomer Tycho Brahe watched the motions of the planet Mars from his observatory on the Baltic island of Hveen; as a result Johannes Kepler was able to show that Mars and the earth and the other planets move in ellipses about the sun. Then the great Isaac Newton proposed his universal law of gravitation and laws of motion, and from these it was possible to derive an exact description of the entire solar system. This occupied the minds of some of the greatest scientists and mathematicians in the centuries that followed.

Unfortunately it is a far more difficult problem to describe the origin of the solar system than the motion of its parts. The materials that we find in the earth and the sun must originally have been in a rather different condition. An understanding of the process by which these materials were assembled requires the knowledge of many new concepts of science such as the molecular theory of gases, thermodynamics, radioactivity and quantum theory. It is not surprising

GREAT FURROW (*right center*) on the surface of the moon must have been made by a tough metallic object. The author believes that this was a fragment of a large body that crashed into the moon from the right.

that little progress was made along these lines until the 20th century.

The Earlier Theories

It is widely assumed by well-informed people that the moon came out of the earth, presumably from what is now the Pacific Ocean. This was proposed about 60 years ago by Sir George Darwin. The notion was considered in detail by F. R. Moulton, who concluded that it was not possible. In 1917 it was again considered by Harold Jeffreys, who thought that his analysis indicated the possibility that the moon had been removed from a completely molten earth by tides. In 1931, however, Jeffreys reviewed the subject and concluded that this could not have happened; since then most astronomers have agreed with him.

But although Moulton and Jeffreys showed the improbability of the origin of the moon from the earth, they proposed theories for the origin of the solar system involving the removal of the earth and the other planets from the sun. Together with James Jeans and T. C. Chamberlin they proposed that another star passed near or collided with the sun, and that the loose material resulting from this cosmic encounter later coagulated into planets. This idea of the origin of the solar system has been widely held right up to the present.

The evidence gathered by our great telescopes now tells us that most of the stars in the heavens are pairs or triplets or quadruplets. We have determined the masses of multiple stars by means of Newton's laws of motion and his universal law of gravitation; we have also studied the velocities of these stars by significant changes in their spectra and by actually measuring the motions of nearby examples. We find that the two stars of a pair seldom have exactly the same mass, and that the ratio of the mass of one star to that of the other varies considerably. Gerard P. Kuiper of the University of Chicago concludes that the number of pairs of stars is entirely independent of the ratios of their masses; that is, there is very little probability that one ratio of masses would occur more often than another. In fact, it would appear that there is about as much chance of finding a pair of stars in which one has one-thousandth the mass of the other as there is of finding a pair in which one is 999 thousandths as massive as the other.

Of course it would be very difficult to see a double star in which the secondary was only a thousandth as large as the primary, particularly if the second emitted no light. The sun and Jupiter, the largest of the planets, might be viewed as such a double star: Jupiter weighs about a thousandth as much as the sun, and it shines only by reflected sunlight. Even from the nearest star Jupiter would

CLOUD OF DUST from which the solar system evolved may have developed this intricate pattern of turbulence, suggested by the German physicist C. F. von Weizsäcker. The dust in each eddy gradually coagulated.

be invisible. There is much evidence, however, that a double star such as the sun and Jupiter should occur as a regular event in our galaxy, and the same considerations would seem to indicate that there may be as many as a hundred million solar systems within it. Solar systems are almost certainly commonplace, and not the special things that one might expect from the collision of two stars.

The Dust Cloud Hypothesis

Many years ago E. E. Barnard of the Yerkes Observatory observed certain black spots in front of the great diffuse nebulae that occur throughout our galaxy. Bart J. Bok of Harvard University has investigated these opaque globules of dust and gas; they have about the mass of the sun and about the dimensions of the space between the sun and the nearest star. Lyman Spitzer, Jr., of Princeton University has shown that if large masses of dust and gas exist in space, they should be pushed together by the light of neighboring stars. Eventually, when the dust particles are sufficiently compressed, gravity should collapse the whole mass, and the pressure and temperature in its interior should be enough to start the thermonuclear reaction of a star.

It would seem reasonable to believe that if a star such as the sun resulted from a process of this kind, there might be enough material left over to make a solar system. And if the process was more complex we might even end up with two stars instead of one. Or again we might have triple stars or quadruple stars. Theories along this line are more plausible to us today than the hypothesis that the planets were in some way removed from the sun after its formation had been completed. In my opinion the older hypotheses were unsatisfactory because they attempted to account for the origin of the planets without accounting for the origin of the sun. When we try to specify how the sun was formed, we immediately find ways in which the material that now comprises the planets may have remained outside of it.

One piece of evidence that must be included in any theory about the origin of the solar system consists in our observation of the angular momentum that resides in the spinning sun and the planets that travel around it. The angular momentum of a planet is equal to its mass times its velocity times its distance from the sun. Jupiter possesses the largest fraction of the angular momentum in the solar system; only about two per cent resides in the sun. Another fact that must

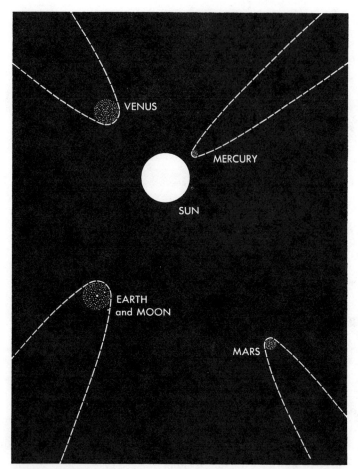

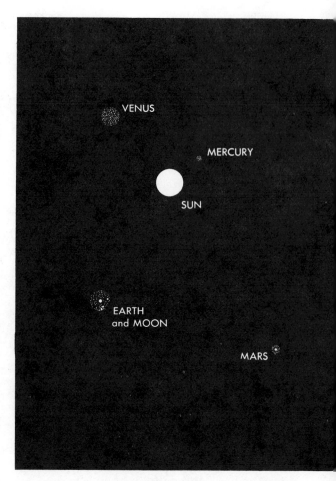

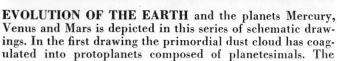

EVOLUTION OF THE EARTH and the planets Mercury, Venus and Mars is depicted in this series of schematic drawings. In the first drawing the primordial dust cloud has coagulated into protoplanets composed of planetesimals. The

gases that have coagulated with the planetesimals are driven away (*dotted lines*) by the pressure of light from the sun. In the second drawing the gas has been completely removed from the proto-

be encompassed by any theory is the so-called Titus-Bode law, which points out in a simple mathematical way how the distances of the planets from the sun vary: the inner planets are closer together and the outer ones are farther apart. This is only an approximate law which does not hold very well, and perhaps more emphasis has been put upon it than it deserves. In my own study of the problem I have looked for other evidence regarding the origin of the solar system.

Some 15 years ago Henry Norris Russell of Princeton and Donald H. Menzel of Harvard pointed out that there was a very curious relationship between the proportions of the elements in the atmosphere of the earth and the atmospheres of the stars, including the sun. It is particularly noteworthy that neon, the gas that we use in electric signs, is very rare in the atmosphere of the earth but is comparatively abundant in the stars. Russell and Menzel concluded that neon, which forms no chemical compounds, escaped from the earth during a hot early period in its history, together with all of the water and other volatile materials that constituted its atmosphere

at that time. The present atmosphere and oceans, they proposed, have been produced by the escape of nitrogen, carbon and water from the interior of the earth. The German physicist C. F. von Weizsäcker similarly suggested that the argon of the air has resulted mostly from the decay of radioactive potassium during geologic time, and has escaped from the interior of the earth. F. W. Aston of Cambridge University also pointed out that the other inert gases, krypton and xenon, were virtually missing from the earth.

The Chemical Approach

My own studies in the origin of the earth started with such thoughts about the loss of volatile chemical elements from the earth's surface. Exactly how did these elements escape from the earth, and when? I came to the conclusion that it was impossible that they were evaporated from a completely formed earth; the evaporation must have occurred at some earlier time in the earth's history. Once the earth was formed its gravitational field was much too strong for volatile gases to escape

into space. But if these gases escaped from the earth at an earlier stage, what is the origin of those that we find on the earth today? Water, for example, would have tended to escape with neon, yet now it forms oceans. The answer seems to be that the chemical properties of water are such that it does not enter into volatile combinations at low temperatures. Thus if the earth had been even cooler than it is today, it might have retained some water in its interior that could have emerged later. But meteorites contain graphite and iron carbide, which require high-temperatures for their formation. If the earth and the other planets were cool, how did these chemical combinations come about?

Indeed, what was the process by which the earth and other planets were formed? None of us was there at the time, and any suggestions that I may make can hardly be considered as certainly true. The most that can be done is to outline a possible course of events which does not contradict physical laws and observed facts. For the present we cannot deduce by rigorous mathematical methods the exact history that began with a globule of dust. And if we cannot

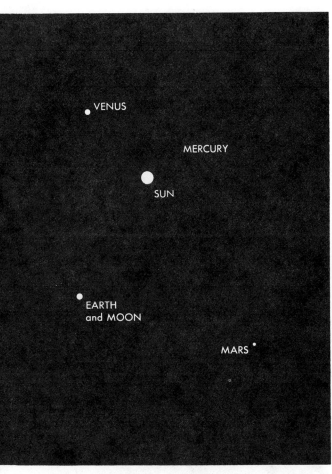

planets. In the third drawing the planetesimals have formed the planets. The relative sizes of the sun and planets and the distances between them have been distorted for purposes of diagrammatic clarity.

do this, we cannot rigorously include or exclude the various steps that have been proposed to account for the evolution of the planets. However, we may be able to show which steps are probable and which improbable.

Kuiper believes that the original mass of dust and gas became differentiated into one portion that formed the sun and others that eventually became the planets. The precursors of the so-called terrestrial planets—Mercury, Venus, the earth and Mars—lost their gases. The giant planets Jupiter and Saturn retained the gases, even most of their exceedingly volatile hydrogen and helium. Uranus and Neptune lost much of their hydrogen, helium, methane and neon, but retained water and ammonia and less volatile materials. All this checks with the present densities of the planets.

It seems reasonably certain that water and ammonia and hydrocarbons such as methane condensed in solid or liquid form in parts of these protoplanets. The dust must have coagulated in vast snowstorms that extended over regions as great as those between the planets of today. After a time substantial objects consisting of water, ammonia, hydro-

carbons and iron or iron oxide were formed. Some of these planetesimals must have been as big as the moon; indeed, the moon may have originated in this way. The accumulation of a body as large as the moon would have generated enough heat to evaporate its volatile substances, but a smaller body would have held them. Most of the smaller bodies doubtless fell into the larger; Deimos and Phobos, the two tiny moons of Mars, may be the survivors of such small bodies.

Massive chunks of iron must also have been formed. On the moon there is a huge plain called Mare Imbrium; it is encircled by mountains gashed by several long grooves. It would seem that the whole formation was created by the fall of a body perhaps 60 miles in diameter; this has been suggested by Robert S. Dietz of the U. S. Naval Electronics Laboratory, and by Ralph B. Baldwin, the author of a book entitled *The Face of the Moon*. The grooves must have been cut by fragments of some very strong material, presumably an alloy of iron and nickel, that were imbedded in this body. Of course large objects of iron still float through interplanetary space; occasionally one of them crashes into the earth as a meteorite.

How were such metallic objects made from the fine material of the primordial dust cloud? In addition to dust the planetesimals contained large amounts of gas, mostly hydrogen. I suggest that the compression of the gases in a contracting planetesimal generated high temperatures that melted silicates, the compounds that today form much of the earth's rocky crust. The same high temperatures, in the presence of hydrogen, reduced iron oxide to iron. The molten iron sank through the silicates and accumulated in large pools.

It now seems that the meteorites were once part of a minor planet that traveled around the sun between the orbits of Mars and Jupiter. The pools of iron that formed in this body may have been a few yards thick. In the case of the object that was responsible for Mare Imbrium and its surrounding grooves, the

depth of the pools must have been several miles. If the temperature of such a planetesimal had been high enough, its silicates would have evaporated, leaving it rich in metallic iron. The object must eventually have cooled off, for otherwise its nickel-iron fragments could scarcely have been hard enough to plow 50-mile grooves on the surface of the moon.

It was at this stage that the planetesimals lost their gases; Kuiper believes that they were probably driven off by the pressure of light from the sun. This left the iron-rich bodies that are today the earth and the other planets. The whole process bequeathed a few meaningful fossils to the modern solar system: the meteorites and the surface of the moon, and perhaps the moons of Mars.

The Moment of Inertia

Recently we have redetermined the density of the various planets and the moon. The densities of some, calculated at low pressures, are as follows: Mercury, 5; Venus, 4.4; the earth, 4.4; Mars, 3.96, and the moon, 3.31. The variation is most plausibly explained by a difference in the iron content of these bodies. And this in turn is most plausibly explained by a difference in the amount of silicate that had evaporated from them. Obviously a planet that had lost much of its silicate would have proportionately more iron than one that had lost less.

It is assumed by practically everyone that the earth was completely molten when it was formed, and that the iron sank to the center of the earth at that time. This idea, like the conception of an earth torn out of the sun, and a moon torn out of the earth, almost has the validity of folklore. Was the earth really liquid in the beginning? N. L. Bowen and other geologists at the Rancho Santa Fe Conference of the National Academy of Sciences in January, 1950, did not think so. They argued that if the earth had been liquid we should expect to find less iron and more silica in its outer parts.

There is other evidence. Mars, which should resemble the earth in some respects, contains about 30 per cent of iron and nickel by weight, and yet we have learned by astronomical means that the chemical composition of Mars is nearly uniform throughout. If this is the case, Mars could never have been molten. The scars on the face of the moon indicate that at the terminal stages of its formation metallic nickel-iron was falling on its surface. The same nickel-iron must have fallen on the earth, but there it would have been vaporized by the energy of its fall into a much larger body. Even so, if the earth had not been molten at the time, some of the nickel-

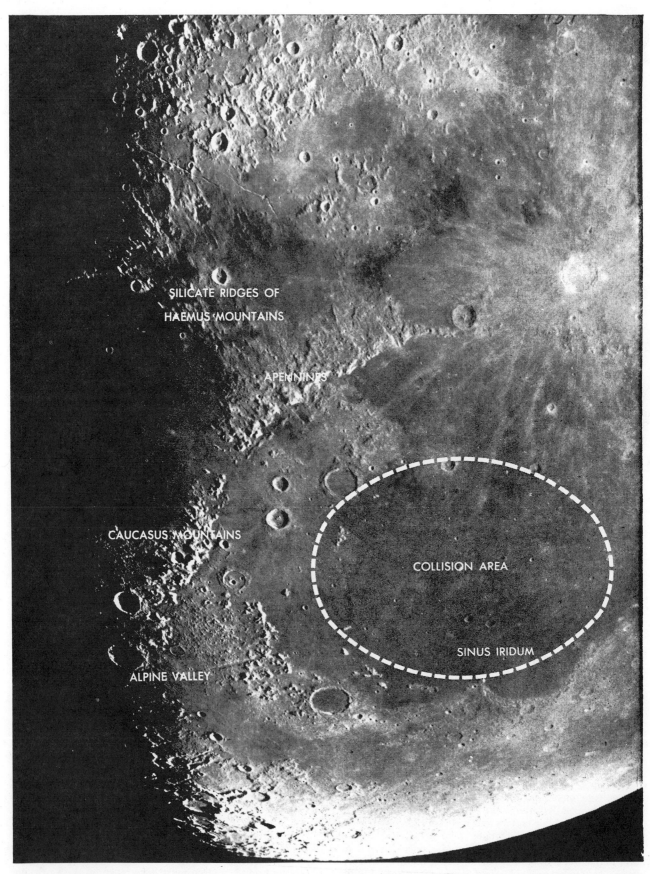

SILICATE RIDGES OF
HAEMUS MOUNTAINS

APENNINES

CAUCASUS MOUNTAINS

COLLISION AREA

SINUS IRIDUM

ALPINE VALLEY

MARE IMBRIUM, a large circular plain near the northern edge of the moon, was probably made by the fall of a body about 60 miles in diameter. This planetesimal apparently ploughed in through Sinus Iridum and spread out in a collision area outlined by relatively small, iceberg-like masses. The rocky silicates of the planetesimal splashed out in the region of the Haemus and the Apennine Mountains. The Alpine Valley, which is the same formation that appears on the preceding page, may have been made by a metallic object.

iron might still be found in its outer mantle.

If there is iron in the mantle of the earth, it may be sifting toward the center of the earth; and if it is moving toward the center of the earth, it will change the moment of inertia of the earth. The moment of inertia may be defined as the sum of the mass at each point in the earth multiplied by the square of the distance to the axis of rotation, and added up for the whole earth. If iron were flowing toward the interior of the earth, this quantity should decrease. It is a requirement of mechanics that if the moment of inertia of a rotating body decreases, its speed of rotation must increase. Finally if the speed of the earth's rotation is increasing, our days should slowly be getting shorter.

Now we know that our unit of time is changing; but it is getting longer, not shorter. That is, the earth is not speeding up but is slowing down. Very precise astronomical measurements, some of them dating back to the observation of eclipses 2,500 years ago, indicate that the day is increasing in length by about one- or two-thousandths of a second per day per century. It has been thought that the lengthening of the day was due to the friction of the tides caused by the sun and the moon. But if we attempt to predict changes in the apparent position of the moon on the basis of this effect alone, we find that our calculations do not agree with the observations at all. If on the other hand we assume that iron is sinking to the core of the earth, the changing moment of inertia would also influence the length of the day. Indeed, calculations made on the basis of both the tides and the changing moment of inertia do agree with the observations.

In order to make the calculations agree we must postulate a flow of 50,000 tons of iron from the mantle to the core of the earth every second. Staggering though this flow may seem, it would take 500 million years to form the metallic core of the earth. Some calculations indicate that it may have taken as long as two billion years. The important thing is that the order of magnitude approaches that of the age of the earth, which is generally given as two to three billion years. If this reasoning is correct, the earth was made initially with some iron in its exterior parts, and it could not have been completely molten.

To complicate matters Walter H. Munk and Roger Revelle of the Scripps Institution of Oceanography have shown that the moment of inertia of the earth is probably decreasing because water is slowly being transferred from the oceans to the ice caps of Greenland and Antarctica, and that this process can account for the lengthening of the day without assuming that iron is moving to the center of the earth, at least not so rapidly as I have calculated. In view of the argument of Munk and Revelle we really have no evidence for the flow of iron to the center of the earth. However, we have little evidence to the contrary. New observations are needed.

The Last Stages

Let us briefly retell what the course of events may have been. A vast cloud of dust and gas in an empty region of our galaxy was compressed by starlight. Later gravitational forces accelerated the accumulation process. In some way which is not yet clear the sun was formed, and produced light and heat much as it does today. Around the sun wheeled a cloud of dust and gas which broke up into turbulent eddies and formed protoplanets, one for each of the planets and probably one for each of the larger asteroids between Mars and Jupiter. At this stage in the process the accumulation of large planetesimals took place through the condensation of water and ammonia. Among these was a rather large planetesimal which made up the main body of the moon; there was also a larger one that eventually formed the earth. The temperature of the planetesimals at first was low, but later rose high enough to melt iron. In the low-temperature stage water accumulated in these objects, and at the high-temperature stage carbon was captured as graphite and iron carbide. Now the gases escaped, and the planetesimals combined by collision.

So, perhaps, the earth was formed! But what has happened since then? Many things, of course, among them the evolution of the earth's atmosphere. At the time of its completion as a solid body, the earth very likely had an atmosphere of water vapor, nitrogen, methane, some hydrogen and small amounts of other gases. J. H. J. Poole of the University of Dublin has made the fundamental suggestion that the escape of hydrogen from the earth led to its oxidizing atmosphere. The hydrogen of methane (CH_4) and ammonia (NH_3) might slowly have escaped, leaving nitrogen, carbon dioxide, water and free oxygen. I believe this took place, but many other molecules containing hydrogen, carbon, nitrogen and oxygen must have appeared before free oxygen. Finally life evolved, and photosynthesis, that basic process by which plants convert carbon dioxide and water into foodstuffs and oxygen. Then began the development of the oxidizing atmosphere as we know it today. And the physical and chemical evolution of the earth and its atmosphere is continuing even now.

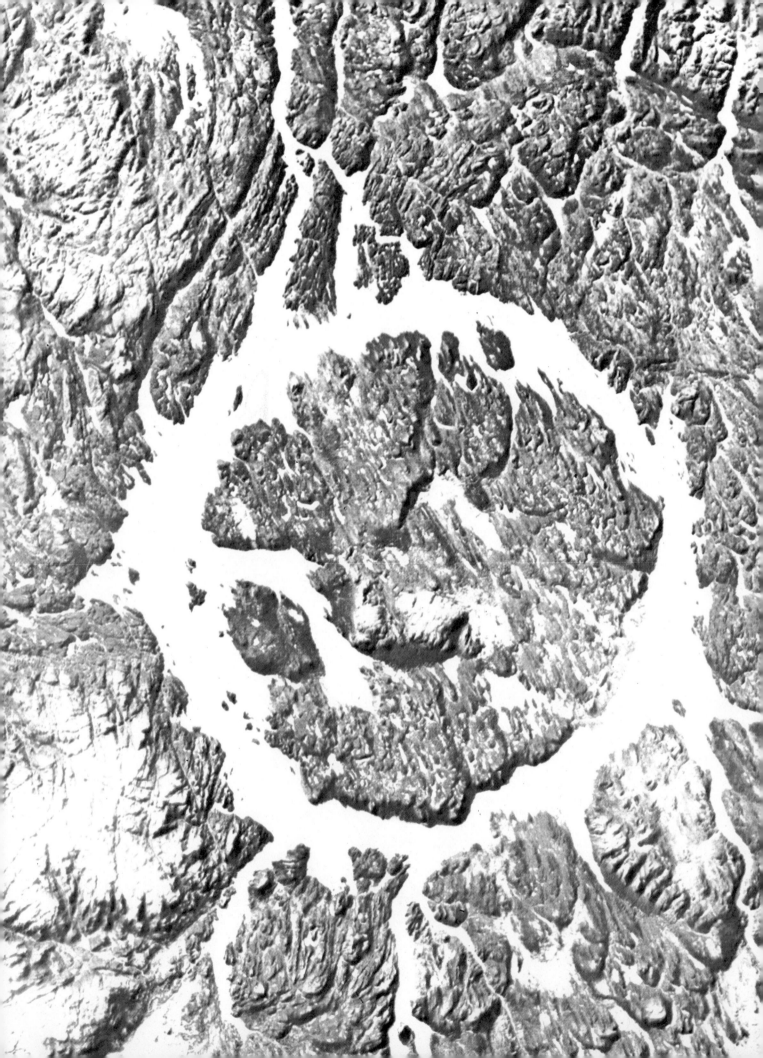

The Earth

3

by Raymond Siever
September 1975

The outstanding feature of our own planet is the dynamic activity of its atmosphere and its crust. Both have been substantially altered by the evolution of living organisms

Apollo astronauts have said that the earth, with its blue water and white clouds, was by far the most inviting object they could see in the sky when they were on the moon. Their bias is understandable. They knew from intimate observation what this planet is like and could translate the sight of clouds, oceans and continents into everyday experience—of, say, a sea breeze blowing surf onto a sunny beach.

Probably the thing people like most about the earth, even if they have not put the thought into words, is its pattern of constant movement. On the earth stillness is remarkable for its rarity. Motion extends from the constant shifting of grains in a sand dune and the movement of bacteria and all other forms of life to the ponderous motions of the entire earth as it vibrates during and after an earthquake.

This planet is active. Indeed, it has been active for 4.6 billion years, and it shows no signs of calming down. The earth's atmosphere, oceans, thin crust and deep interior have been in motion since they were formed. Life has been an integral part of the surface for at least four-fifths of the planet's history.

As a consequence of its steady activity over this long span of time the earth has evolved through a series of quite different stages, maintaining during the entire time a state of dynamic equilibri-um. The balance involves an exchange of matter and energy between the interior, the surface, the atmosphere and the oceans. It also involves sharing the radiation of the sun with the other members of the solar system. The study of geology, aided by work in geochemistry, geophysics and paleontology, has shown how the earth's surface skin has evolved. That knowledge, coupled with a firm theory about the constitution of the earth's interior and certain hypotheses on how the interior moves, provides a means for constructing a theory of how the planet evolved.

The article by A. G. W. Cameron [see "The Origin and Evolution of the Solar System," page 8] describes the origin of the earth and the other planets through the condensation of particular regions of the solar nebula. The original composition of the nebula and its later composition are deduced from the composition of the earth's rocks, of rocks brought back from the moon, of meteorites and of the atmospheres of the earth, Mars, Venus and Jupiter.

The theory of the earth's growth most favored by people who have studied the subject infers a gradual condensation and accretion of a solid planet as it swept up enormous quantities of small particles from the nebular disk that gave rise to the present solar system. As the planet grew it began to heat up as a result of the combined effect of the gravitational infall of its mass, the impact of meteorites and the heat from the radioactive decay of uranium, thorium and potassium. (Although potassium is not normally regarded as being radioactive, .01 percent of the element on the earth is the radioactive isotope potassium 40.) Eventually the interior became molten. The consequence of the melting was what has been called the iron catastrophe, involving a vast reorganization of the entire body of the planet. Molten drops of iron and associated elements sank to the center of the earth and there formed a molten core that remains largely liquid today.

As the heavy metals sank to the core the lighter "slag" floated to the top—to the outer layers that are now termed the upper mantle and the crust. Accompanying the rise of the lighter elements, such as aluminum and silicon and two of the alkali metals, sodium and potassium, were the radioactive heavy elements uranium and thorium. The explanation for the rise of these heavy elements lies in the way atoms of uranium and thorium form crystalline compounds. The size and chemical affinities of the atoms prevent them from being accommodated in the dense, tight crystal structures that are stable at the high pressures of the deep interior of the earth. Therefore the uranium and thorium atoms were "squeezed out" and forced to migrate upward to the region of the upper mantle and the crust, where they could fit easily into the more open crystalline structures of the silicates and oxides found in crustal rocks.

As the earth was differentiating into a core, a mantle and a crust, the material at the top was also splitting into different fractions. The lower parts of the crust are composed of basalt and gabbro,

CRATER ON THE CANADIAN SHIELD, seen in the satellite photograph on the opposite page, was formed more than 200 million years ago by the impact of a large meteorite. The impact scar, a ring-shaped formation about 60 kilometers (37 miles) in diameter, is filled today by Lake Manicouagan, a major reservoir in northeastern Quebec. The accumulation of snow on the lake ice enhances the visibility of the formation. The bedrock, a Precambrian anorthosite, was melted and shattered by the impact; a peak of shocked rock remains in the center of the formation. Unlike the impact craters on Mercury and the moon, most of the craters on the earth were long ago erased or buried by erosional or tectonic processes. Many craters on the Canadian Shield, however, were preserved under layers of sediment and reexposed by glacial action. In this area glaciation removed 100 meters of sedimentary cover.

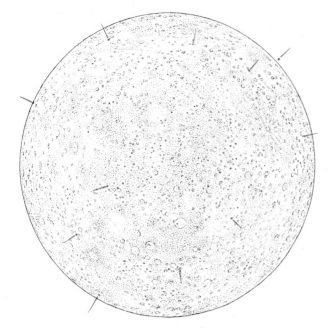
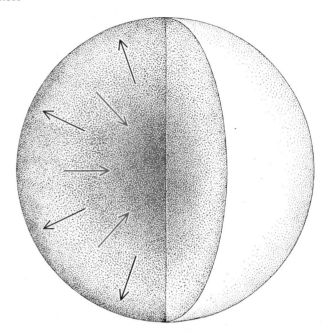

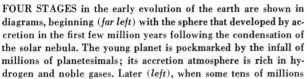

FOUR STAGES in the early evolution of the earth are shown in diagrams, beginning (*far left*) with the sphere that developed by accretion in the first few million years following the condensation of the solar nebula. The young planet is pockmarked by the infall of millions of planetesimals; its accretion atmosphere is rich in hydrogen and noble gases. Later (*left*), when some tens of millions of years have passed, a combination of gravitational compression, radioactive decay and impact heating produces melting and differentiation; heavy core and mantle materials sink inward and light crustal materials float outward. The solar wind sweeps away the accretion atmosphere; it is replaced by a primordial atmosphere rich in methane, ammonia and water. The Archaean era (*right*), be-

dark rocks containing calcium, magnesium and iron-rich compounds, mainly silicates. They were derived by partial melting and differentiation of the denser materials of the upper mantle. The basalt and the gabbro were themselves differentiated by fractional crystallization and partial melting and so, as lighter fluids, were pushed up through the crust. In the upper layers of the crust and at the surface they solidified to form the lighter igneous rocks such as granite, which are enriched in silicon, aluminum and potassium.

The question of how much of this "sweating out" process was completed at the early stage I am describing remains unresolved. Some geologists argue that a large amount, perhaps the bulk, of the granitic crust had already formed by this stage. Others cite the possibility that the process may hardly have begun even a billion years after the formation of the earth.

One result of the heating up of the interior was the inception of volcanic activity and mountain building. They contributed not only to the shaping of the surface but also to the immense change in the composition of the interior. During that time various gases, which had been locked in the materials of the planet when those materials originally accreted, began to find their way to the surface. They included carbon dioxide, methane, water and gases con-

taining sulfur. The gases must have leaked to the surface in tremendous volume during the period of reorganization and differentiation. At the surface they stayed, since the earth's gravity was strong enough to prevent all but the lightest elements (hydrogen and helium) from escaping into space. The temperature at the time must have been low enough to allow the condensation of water. Dissolved in that water, the other gases combined chemically with elements such as calcium and magnesium, which were leached from surface rocks as rains began to weather them. If the temperature had been higher, the effect of the dense atmosphere with its large content of carbon dioxide would have been to institute the kind of "greenhouse effect" that seems to have arisen on Venus, producing that planet's hot, cloudy atmosphere.

As the surface of the earth cooled and the oceans formed from the condensation of water, the processes of erosion by wind and water began to operate in much the same way that they do today. Liquid water became the dominant mode for transporting and redistributing the debris of eroding mountains. The river systems of the surface are the visible traces of the networks that carry eroded material to the oceans, where much of it accumulates as aprons of sediment along the continental shelves and continental rises. The rest of the sedi-

ment is spread as thin layers over the ocean deeps by slow settling and the motions of turbidity currents.

A number of thoughtful geochemists and geophysicists have speculated on a somewhat different chain of events leading to the early accretion of the earth from the condensing solar nebula. According to these views, the earth and the other planets are the products of a gradual condensation of the solar nebula during which certain of the heavy elements, mainly iron, crystallized first, while the lighter fractions of the nebula were still in gaseous form. In that process the core of the accreting planet would have been iron-rich in the first place, with successively lighter fractions, roughly corresponding to the order of their crystallization from a gas, being accreted on the outside as the planet grew.

Whatever the mechanism of accretion, the story of the earth's later evolution (after the first billion years) is largely told by the record contained in the rocks of the crust. What they reveal is best told in terms of a geologic "clock" that began to run in Precambrian times. The oldest rocks now known are a series of metamorphosed sedimentary and volcanic rocks, which from their content of radioactive elements can be given an age of about 3.7 billion years. They are

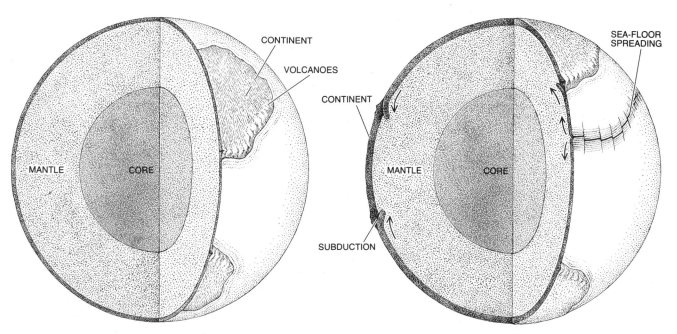

CONTINENT

VOLCANOES

CONTINENT

MANTLE   CORE

SUBDUCTION

SEA-FLOOR SPREADING

CONTINENT

MANTLE   CORE

tween 3.7 and 2.2 billion years ago, is marked by cooling. As atmospheric water and gases condense, the oceans appear; the earliest continents arise and volcanic activity is intense. Chemical reactions between water, gases and crustal materials give rise to sediments and solutions. The Proterozoic era (*far right*) follows the Archaean era. It harbors a few scant traces of plant and animal life. It ends some 600 million years ago as the Paleozoic era, with its rich fossil record, begins. During that era's 1.5-billion-year span crustal cooling and thickening continue. Crustal plates begin migrating as such mechanisms as sea-floor spreading and subduction become established. Long before the end of the Proterozoic era both the interior and the surface of the planet have assumed their modern character.

much older than most of the very old rocks from the interval of time known to geologists as the Archaean. The rocks of that time are roughly defined as being older than 2.2 to 2.8 billion years. (The age of the boundary with younger eras varies in different parts of the earth's ancient rock terrains.) Much of the rock record is fragmentary, but it is tangible, and one no longer has to rely solely on plausible theory.

The Archaean rocks appear to be somewhat different from the rocks of all succeeding eras in the sense that certain rock types are abundant almost to the exclusion of many other types commonly found later. Archaean rock series tend to be dominated by basalts and andesites, which are volcanic rocks rich in iron and magnesium, deficient in sodium and potassium and relatively low in silica. The sandstones and shales of Archaean time were derived by the weathering and reworking of those volcanic rocks. Large bodies of granite—rocks richer in alkalis and silica—are absent. Such a skewed composition with respect to later rocks suggests that the sweating out of granitic rocks by fractional crystallization and partial melting of less silicic rocks was not as advanced as it became later.

The Archaean rocks also suggest that the tectonic style of the time, that is, the mountain-making activity by which the surface was shaped, differed from the pattern of today. The present theory of plate tectonics visualizes large plates of the lithosphere (which includes the crust and part of the upper mantle) moving laterally over the asthenosphere (a hot, plastic and perhaps partly molten layer of the mantle). The driving force is movement in the mantle, although the precise nature of that movement is uncertain. The geologic activity of earthquakes, volcanoes and mountain building is concentrated along the plate boundaries.

Granted that the Archaean rocks are widely dispersed and offer only a few bits of information, the study of the oldest terrains of Archaean areas in Canada and areas of similar age in Africa and Scandinavia does not suggest mountain building along the boundaries of large plates. It does suggest patterns of intense deformation along the boundaries of irregular areas of far smaller extent than plates. Many geologists suspect that the Archaean was a time of very thin lithospheric crust, extensive volcanism and some jostling movement of many small, thin "platelets," with "sutures," or crumpled deformational belts, welding them together.

Although the Archaean era differed markedly from the present in tectonic style and in the average composition of its volcanic rocks, it was the same as the present in all essential processes of erosion and sedimentation on the surface.

All the earmarks of weathering, mechanical breakup of rocks, transportation by rivers and sedimentation in regions where the crust gradually subsided and allowed great thicknesses of sediment to accumulate are found in Archaean sediments, as was pointed out more than 30 years ago by Francis J. Pettijohn of Johns Hopkins University, who was studying early Precambrian sedimentary rocks in the region of Lake Superior. Looking at those sandstones, shales and conglomerates, it is difficult to see any significant difference between them and more recent ones, all being the hardened equivalents of the gravels, sands and muds of today.

The erosion and chemical decay of rocks today are profoundly affected by the presence of land plants. It is known, however, that the higher (vascular) land plants did not evolve until two billion years after Archaean time, that is, in the middle of the Paleozoic era. Perhaps before the plants evolved, lower forms of life existed on the land, as they surely did in the sea.

Evidence of algal life late in the Precambrian era was obtained some years ago when the paleobotanist Elso S. Barghoorn of Harvard University, working with the late Stanley A. Tyler, a sedimentologist at the University of Wisconsin, discovered microscopic remains of algal organisms in the Gunflint

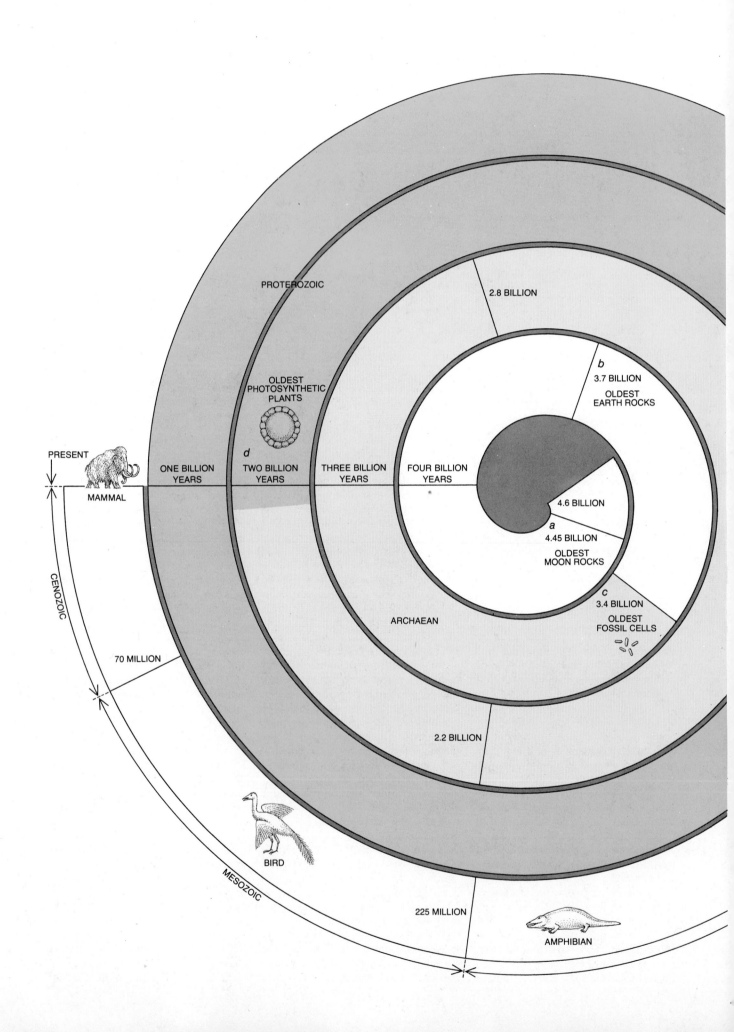

PROTEROZOIC

2.8 BILLION

b
3.7 BILLION
OLDEST
EARTH ROCKS

OLDEST
PHOTOSYNTHETIC
PLANTS

d

PRESENT

ONE BILLION
YEARS

TWO BILLION
YEARS

THREE BILLION
YEARS

FOUR BILLION
YEARS

4.6 BILLION

MAMMAL

a
4.45 BILLION
OLDEST
MOON ROCKS

CENOZOIC

c
3.4 BILLION
OLDEST
FOSSIL CELLS

ARCHAEAN

70 MILLION

2.2 BILLION

MESOZOIC

BIRD

225 MILLION

AMPHIBIAN

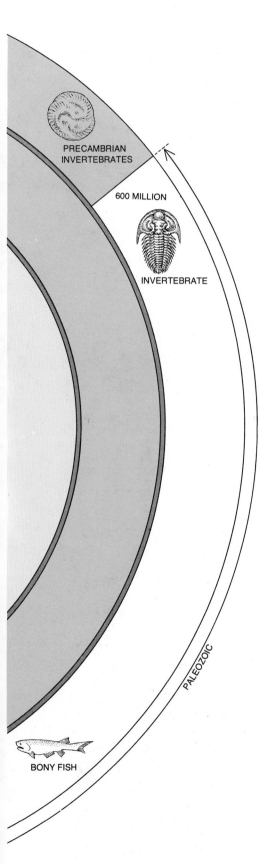

chert, a dense sedimentary rock made of silica. The Gunflint chert has been dated by its content of radioactive elements and their decay products to an age of about two billion years. Since then other organic structures that look like the remains of organisms have been found in even older rocks. The oldest of them, aged about 3.4 billion years, is the Fig Tree chert of Swaziland in Africa.

This kind of search for evidence of ancient life is a painstaking, laborious process. Thousands of rock specimens have to be sawed into ultrathin slices and then polished so that they can be studied under the light microscope and the electron microscope. Although organic carbon had been found in old rocks long before the discovery of the Gunflint and Fig Tree cherts, one could always hypothesize a variety of ingenious chemical mechanisms to account for it. The more recent evidence of distinctive forms of cellular life in ancient times is difficult to refute.

How life began on the earth is another story. It is the story of plausible chemical mechanisms that can be deduced from certain assumptions about the early chemical environment of the surface. One begins by inferring an early Archaean atmosphere (which had been built up by the escape of gas from the interior) that was dominated by water, methane and ammonia. Free oxygen was absent, since free oxygen is a product of life and not an antecedent to it. The atmosphere may also have contained appreciable quantities of carbon dioxide.

The existence and character of this atmosphere are related to the fact that the earth is smaller than Jupiter and larger than the moon. Jupiter was able to hold its hydrogen, which was by far the most abundant element in the solar nebula. The moon could not hold any of its gas.

In the earth's envelope of air and below it, in the surface waters of the sea and in large lakes, ultraviolet radiation from the sun was intense. The surface was not screened from the ultraviolet by a layer of ozone, as it is now, for

want of the oxygen (O_2) from which ozone (O_3) is derived. The high energy of the ultraviolet radiation promoted the synthesis of a variety of organic compounds, for example amino acids. Perhaps many of these compounds were already here, since it is now known that a number of simple organic compounds are present in interstellar space.

The synthesis of transitory organic compounds, however, is not the same as making life. The next steps had to be the growth of large molecules and, before long, the growth of the nucleic acids that would eventually provide the genetic mechanism of reproduction so that cells could divide and give rise to new cells like themselves.

One cannot be sure of the range of chemical environments that will support life. (The uncertainty may be diminished by the U.S. spacecraft scheduled to land on Mars next year.) All that is known now is that the earth supports life and that its life depends on the continuous existence of liquid water. At present the earth is the only planet known to satisfy that condition. The earth's continuous record of life for at least the past 3.5 billion years shows that liquid water has been available during all that time.

Once life evolved, it began to exert an important effect on the surface of the earth and the gaseous envelope surrounding it. In the Bitter Springs formation of central Australia, which is a little less than a billion years old, paleobotanists have found cellular algae showing many of the geometric characteristics of the blue-green algae that today, like all other photosynthetic plants, evolve oxygen as a waste product. By the end of the Proterozoic era, which lies between Archaean time and the beginning of the Paleozoic era, there must have been enough oxygen in the atmosphere to support the evolution of higher organisms. They were the metazoans—animal organisms having many cells with differentiated characteristics. All these organisms appear to need at least small

SPIRAL "CLOCK" shows the passage of 4.6 billion years of earth history; each revolution of the hand takes a billion years. Moving clockwise toward zero (the present), the hand passes the first significant datum at 4.5 billion years before the present (a); this is the age of the oldest moon rocks known. The first complete revolution of the hand brings it to the oldest sedimentary rocks known on the earth, Archaean-era strata in Greenland (b). Some 350 million years later (c) the hand passes the earliest of certain fossil-like microstructures found in rocks from Swaziland in Africa; these may represent the planet's first flora. Almost a revolution and a half more are required to carry the hand past the algalike plants, roughly two billion years old (d), that are found in Canadian chert. Only two more revolutions remain to go; the Archaean era lies behind and most of the Proterozoic lies ahead. Half a revolution more will pass by three more familiar divisions of geologic time: Paleozoic, Mesozoic and Cenozoic. Human history occupies a hair's breadth to the left of zero.

quantities of free oxygen for their biochemical processes.

Oxygen is not the only atmospheric gas that comes from life. Methane, for example, is present in minute quantities. Its source seems to be primarily the methane-producing bacteria that yield the abundant "marsh gas" over swamps. The atmosphere also contains other gases that are the distinctive product of life rather than of simpler nonbiological chemical reactions.

The Proterozoic era was a time when the world was populated by bacteria, algae and other primitive single-cell organisms, probably on land as well as in the sea. Their influence on surface processes is seen in the Proterozoic rocks. It is most distinctive in stromatolites: rock formations consisting of the limy secretions of mats of filamentous algae and the sediment trapped by them. Stromatolites are known today in places such as the Bahamas and Bermuda, where limestone is being laid down in tidal flats. Other evidence of Proterozoic life is found in the existence of a few coal beds formed from masses of carbonized algal remains.

If an observer had looked down on the earth from an artificial satellite in Proterozoic time, he would have described the surface in much the same way that an observer similarly situated would describe it today. Only a sensor that could determine the chemical composition of the atmosphere would reveal any differences. The evidence for the similarities is in the Proterozoic rocks, which are of the same types and abundances as the rocks from all later ages.

By late Proterozoic time the earth-moon system, after early instabilities, had settled down into much the same system we see today. Tides would have been somewhat higher than they are now, but they would not have been grossly different. At about the time the moon became a cold planet the long heating and differentiation of the earth's upper mantle and crust resulted in the extensive intrusion of great bodies of granitic rock and in patterns of mountain belts that suggest a plate-tectonic origin.

Another kind of evidence from both Proterozoic rocks and more recent ones reveals periodic reversals of the earth's magnetic poles during much of the earth's history. As a heated rock cools it is magnetized in the direction of the earth's magnetic field, and the pattern is frozen in when the rock solidifies. Similarly, certain sediments that contain magnetic particles will record the direction of the field at the time they were deposited. The causes of reversals lie in instabilities in the fluid motion of the core, which is the driving force that creates the earth's magnetic field.

The same kind of paleomagnetic evidence reveals what has been called polar wandering, although it is not that the North and South poles have moved but rather that the surface features of the earth have shifted in relation to the poles. The conclusion is reinforced by paleoclimatic evidence, that is, the geologic record of ancient climates, such as the occurrence of coal beds in polar regions and glacial deposits near the Equator. From this kind of information it appears that a major continent was at the South Pole in Proterozoic time and that continental drift was already established as a major process affecting paleogeography.

The rocks also record from this time a major glacial epoch, the first for which there is firm evidence. The evidence is

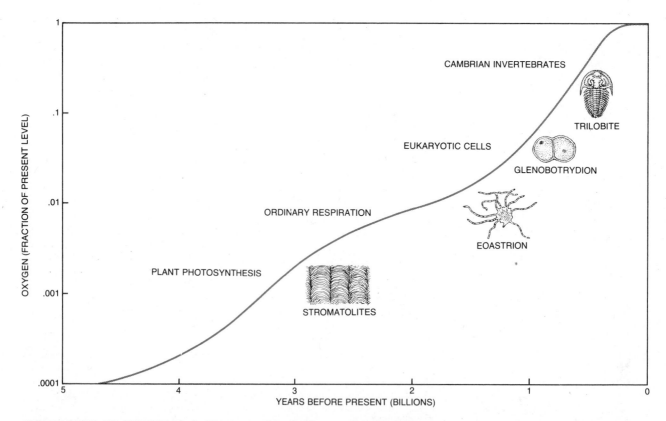

APPEARANCE OF OXYGEN in significant quantities in the earth's atmosphere, a late development, is an event that remains a subject of controversy. One hypothesis is shown in this semilogarithmic graph. The abscissa intervals are billions of years before present; increases in the oxygen supply, from trace amounts to the present quantity (about 20 percent of the atmosphere), are indicated on the ordinate. The process must have been gradual and must also have been related to an increase in the numbers of photosynthetic plants. The oxygen level may have risen to 10 percent of its present value a billion years ago but no evidence of oxygen-dependent animal life in any abundance appears until the time of a steep increase in the oxygen supply at the end of the Proterozoic era.

insufficient to reveal the details of that ice age—whether it was of the same extent as the recent (Pleistocene) ice ages and whether, like them, it had many episodes of glacial advance and retreat. One can only assume that the mechanisms postulated for the Pleistocene ice ages are general ones that are set in motion once a continental land mass lies at one of the poles and restricts the ability of the ocean and the atmosphere to distribute heat evenly around the globe. To an external observer at that time the earth would have looked a little like Mars, except that there were still oceans at the Equator. One of the interesting questions about the earth's glacial epochs is why the earth remained poised at a temperature distribution sufficiently low to give rise to large polar ice caps but not to a complete freeze-over.

Just as human history merges with pre-history, so the most recent 570 million years of the earth's history (starting with the Paleozoic era) connect with the nine-tenths of earlier evolution that were long thought to be a mystery. For more than a century the past 570 million years have been regarded as the geologically "known" period; it is therefore often called the Phanerozoic from the Greek *phaneros*, "to reveal." Although early geologists recognized that some Precambrian terrains were mappable by ordinary geologic methods, it was the absence of fossils having recognizable affinities with forms of the present that made it unknowable. The stratigraphic time scale, a marvelously detailed and precise clock, depends on the rapid evolutionary changes in higher forms of life that are recorded in the fossilized remains of corals, mollusks and thousands of other kinds of metazoan life.

Students of the earth's history never tire of marveling at the extraordinary speed of the coming of the metazoans. For between three and four billion years, almost its entire history, the earth was populated by single-cell life. Within at most a few hundred million years thereafter a fantastic diversity of invertebrate organisms appeared. All the major phyla of the animal kingdom became established quickly, and the vascular plants and the vertebrates soon followed.

Was all of this an accident, a favorable conjunction of continents, seas and environmental niches? Or was it the inevitable consequence of the buildup of oxygen in the earth's atmosphere by photosynthesizing algae? The best guess now is that it was the evolution of the atmosphere to near its present level of oxygen that stimulated biological inven-

tion. One such invention was animal shell, which served as armor to protect soft bodies from predators and as a base of attachment for muscles. Shells provide the basis for our understanding of the subsequent course of evolution of both the planet and its inhabitants. A paleobiological record based solely on the soft parts of organisms would provide only the dimmest outlines of the past.

The shells are more than markers of history: they influenced important changes in the dynamics of the earth's exterior. The oceans became populated with organisms that secreted calcium carbonate, calcium phosphate and silica in enormous quantities. Their remains were deposited as sediment, ultimately to become limestone, chert and phosphatic limestone or phosphate rock (a major source of agricultural fertilizer).

The more precise knowledge afforded by Paleozoic records enables geologists to trace the effects of continental drift. In particular it is possible to map more confidently the shape of the early Atlantic Ocean that lay between the European-African land mass and the Americas before the supercontinent of Pangaea was assembled at the close of the Paleozoic era. The assembly of Pangaea was one of the rare, special events of the later history of the earth, one of the important perturbations of the otherwise more or less evenly ordered evolution of the planet.

One of the major consequences of the assembly of Pangaea was the extinction of hundreds of species of invertebrates and the beginning of a wholesale change in the kinds and relative populations of the different animal and plant species. Most of the expanse of shallow shelf surrounding each continent disappeared as the continents collided, leaving only one narrow perimeter around the supercontinent. The shelves had harbored the most productive biological populations of the Paleozoic world. The geographic constriction and the concurrent climatic extremes, including the glaciation of parts of what are now Africa, Australia and South America, were enough to decimate many species. The survivors went on to found the new stocks of the post-Paleozoic world.

Pangaea rifted apart in the Triassic period (the earliest part of the Mesozoic era), and with that event and the following opening of the Atlantic Ocean and the drifting of the continents to their present position the story of the earth's physical evolution is largely told. The oldest parts of the ocean floor that are now preserved came into being at this

time, and so began a decipherable history of the world's oceans. It is read from the magnetic "stripes" and the fracture zones of the sea floor formed at mid-ocean ridges and rifting zones.

The new forms of life that evolved early in the Mesozoic era give the appearance of the modern world. Flowering plants appeared, and the lands became covered with the colors of the flowers and foliage of deciduous trees, the grasses and a great number of shrubs and flowers. In the sea new photosynthesizing algae, the diatoms, evolved; they are single-cell creatures secreting thin shells of silica. The diatoms became responsible for much of the primary photosynthetic production of organic matter in the sea.

At about the same time the calcareous foraminifera appeared. They are single-cell animals that live off the plants at the surface of the sea. Their shells of calcium carbonate, raining steadily to the bottom of the oceans, became the source of a new kind of deep-sea sediment, the foraminiferal oozes. The remains of these foraminifera became part of another detective story: the deduction of ancient sea temperatures, and thus of world climates, from the isotopic composition and external form of the shells. Both the shape of a shell and the relative proportions in it of the normal oxygen atom (oxygen 16) and the rare heavy isotope (oxygen 18) reflect the temperature of the water in which the animal lived. The temperature of the oceans as measured in this way has revealed an important climatic change.

Over most of the past 50 million years (during much of the Cenozoic era) the earth was cooling. This culminated in the past few million years in repeated glaciations. The more recent ones have been witnessed by and have affected the evolution of a new species: man. Already advanced on his course of evolution, man in his primitive cultures was displaced as the glaciers covered much of northern Europe, Asia and North America. In the short 10,000 years since the glaciers retreated to their present ice-cap size (probably a temporary retreat) man became the species that spread and occupied almost every environment of the surface of the planet. As he did so he became the latest of the biological populations to profoundly affect the course of the earth's history. He is only now becoming aware that some of his activities may alter the thin envelope of the atmosphere and the oceans and the fresh waters that make his existence possible.

II

PREBIOTIC CHEMISTRY

II   PREBIOTIC CHEMISTRY

INTRODUCTION

Chemical evolution is the term commonly used to describe primeval synthesis of organic compounds from volcanic gases on the newly formed Earth. It is a misnomer—chemicals don't evolve. But it does suggest the vast range of complex reactions that might occur when gases such as carbon monoxide, carbon dioxide, nitrogen, hydrogen, and water vapor are exposed to the enormous and varied sources of energy impinging on the primitive Earth.

John S. Lewis, in "The Chemistry of the Solar System," sets the prebiotic stage. He shows the *why* behind volcanic emissions and relates the chemical elements to the primeval scene.

George Wald, in "The Origin of Life," assumes that a primeval soup of organic molecules somehow originated. He then proceeds to show that *given enough time,* random combinations of molecules might form biologically relevant associations and, ultimately, life. Although stimulating, this article probably represents one of the very few times in his professional life when Wald has been wrong. Examine his main thesis and see. Can we really form a biological cell by waiting for chance combinations of organic compounds? Harold Morowitz, in his book *Energy Flow and Biology,* computed that merely to create a bacterium would require more time than the Universe might ever see if chance combinations of its molecules were the only driving force. As later articles will show, the oldest fossils indicate that the most primitive cells arose early in the history of the Earth. Thus relatively short spans of time were available for the origins of cellular life. The age of the Earth is some 4500 m yr (million years), and the oldest known fossils are dated at about 3000 m yr. Formation of the Earth's core and the development of stable ocean systems and a solid crust required about 500 m yr, leaving a gap of only 1000 m yr. During this early era small organic compounds must accumulate, biological polymers must form, protocells must arise, and a genetic and protein-synthesizing system must evolve. These events are not consistent with the Wald hypothesis of random associations. Since Wald's article was written, biological macromolecules have been found to possess the amazing ability to "self-assemble." The various proteins of a virus, when isolated and then remixed, spontaneously reaggregate to form intact virions. Subunits of enzymes, once separated, will aggregate again to form enzymatically active proteins. The separated two strands of the DNA molecule will reform to generate a duplex indistinguishable from the original. Dissociated membranes of cells will reunite to form boundaries identical to the original. And so on.

All biological macromolecules seem to possess this marvelous property, which we have only recently discovered. This automatic assembly aspect of large molecules makes the spontaneous organization of protocells from primitive polymers more a matter of internal necessity that of chance. If we add this newer knowledge to Wald's argument, we cannot but agree with his conclusions. Is this another example of the essential truth versus the details of a good basic explanation?

C. L. Stong's "Amateur Scientist" article points out how very easily various energy sources can react with simple gases to produce a whole spectrum of biologically relevant organic compounds. The article mirrors the nuts-and-bolts reality of science from the point of view of the chemist attacking the problem of the prebiological chemistry of simple gases.

The theoretical postulates of prebiological chemistry are few and nicely simple. Gases emitted from volcanoes, including carbon monoxide, carbon dioxide, nitrogen, hydrogen, and water vapor, react with a plethora of energy sources—solar radiation, heat, electrical discharges, meteorite impact events, radioisotope decay—to yield a vast range of organic molecules. Contained within this variety are those molecules used in our biochemistry. By a series of selection mechanisms at present unexplained, our subset of biological molecules becomes chosen to appear in evolved biochemistry.

From an observational point of view, this theory might seem a bit contrived. Sugars have never been recovered from chemical evolution reconstruction experiments. Purines and pyrimidines, the core of RNA and DNA molecules, can only be made using highly specialized reactants in unlikely concentrations. The problem that prebiotic chemical events pose to the experimentalist is one of proper simulation. How can we copy in the laboratory the grand panorama of reactions that must have occurred on our primitive planet? The laboratory can only be a small window peering at a scene vast in time and space. By this rationale we excuse the experimental shortcomings of the field.

Sugars, the heart of our biochemical energetic metabolism, are chemically reactive molecules that would combine almost immediately with other products in most simulation experiments. To the organic chemist, it comes as no surprise that these molecules are not found.

Only recently has it been shown by the author that purines and pyrimidines can be formed in spark discharge experiments if and only if the relative amount of hydrogen in the starting gas mixtures is kept low.

At the University of Chicago, Clifford Matthews has shown that peptides might be formed directly from polymers of hydrogen cyanide. S. Akabori, in Japan, has proposed another direct and plausible mechanism for the direct formation of peptides. These insights indicate our progress in resolving the concentration gap problem.

At the NASA Ames Research Center, James Lawless and his associates discovered that certain metal-clay complexes would select for the protein amino acids and aid in their polymerization. These metal-clays also preferentially adsorb nucleotides. Thus the selective role of clays could have played a vital role in a prebiotic chemistry.

In this way research on prebiotic chemistry progresses, continually refining the simulation parameters and adding to the list of the many organic products formed from simple volcanic gases. But the key concept was there at the beginning: a complex set of organic molecules invariably results from the interactions of energy sources and volcanic gases.

Suggested Further Reading

Buvet, René, and Cyril Ponnamperuma (eds.). 1971. *Molecular Evolution I: Chemical Evolution and the Origin of Life.* North-Holland, Amsterdam.
Calvin, Melvin, 1969. *Chemical Evolution.* Oxford University Press, New York.
Fox, Sidney W. (ed.). 1965. *The Origin of Prebiological Systems.* Academic Press, New York.
Hanawalt, P. C., and R. H. Haynes (eds.). 1973. *The Chemical Basis of Life: An Introduction to Molecular and Cell Biology, Readings from Scientific American.* W. H. Freeman and Company, San Francisco.
Morowitz, Harold J. 1968. *Energy Flow in Biology.* Academic Press, New York.

4

The Chemistry
of the Solar System

by John S. Lewis
March 1974

*The sun, the planets and other bodies in the system
formed out of a cloud of dust and gas. What processes
in the cloud could account for the present
composition of these objects?*

The era of space exploration, with the landing of spacecraft on the moon and Venus and the flyby missions to Mars, Venus and Jupiter and now to Mercury and Saturn as well, has greatly enlarged our knowledge of the composition of the planets and satellites and the evolution of the solar system. Our modern understanding of the composition of the solar system nonetheless has a substantial history. It began in the 1930's with the work of Rupert Wildt on the physics and chemistry of the Jovian planets: Jupiter, Saturn, Uranus and Neptune. Wildt's investigations revealed that the atmosphere of Jupiter contains large amounts of the gases ammonia (NH_3) and methane (CH_4) and gave strong reason to suspect that all the Jovian planets consisted largely of hydrogen.

Around 1950 there was a second period of vigorous inquiry into the chemical nature of the solar system. At that time Harrison Brown proposed that the objects in the solar system could be divided into three classes according to their density and their chemical composition. The first class was the rocky objects such as the terrestrial (earthlike) planets, their satellites, the asteroids and the meteoroids. The second class was the objects consisting of both rocky and icy materials such as the nuclei of comets and the satellites of the outer planets. The third class was the sun and the Jovian planets, which consist mainly of matter in the gaseous state. At the same time that Brown suggested these classes Gerard P. Kuiper and Harold C. Urey were working on complex theories of the origin of the solar system. Urey stressed the importance of meteorites as clues to the origin of the planets, and he calculated the stability of the chemical equilibrium of numerous meteoritic minerals. Since that time the detailed study of the rocks brought back from the moon, the spectra obtained of comets and the numerous photographs made from spacecraft of the moon and the planets, together with what we know about the geology of the earth, have all contributed to the current picture of the solar system and its origin.

The Solar Nebula

The solar system is not a homogeneous mixture of chemical elements. Moving outward from the sun, there is a general trend for the abundance of the volatile, or easily evaporated, elements to increase with respect to the abundance of the nonvolatile elements. That general trend appears to be evidence for the hypothesis that the Jovian planets and the other bodies in the outer solar system formed at low temperatures and the terrestrial planets formed at high temperatures. According to this hypothesis, the solar system formed out of the solar nebula: a large cloud of dust and gas. The center of the nebula was hot (several thousand degrees Kelvin) and the edges were cold (a few tens of degrees K.).

The difference in temperature meant that the composition of the dust must have varied radically with distance from the center. Close to the protosun in the center all materials would have been totally evaporated. At a distance of perhaps 20 million miles from the protosun, a fifth of the way to the present orbit of the earth, a very few nonvolatile materials could have condensed into solid particles. From that point out to the present location of the asteroid belt the dust would have been composed primarily of grains of rocky material with only a limited content of volatile substances. Beyond the location of the asteroid belt the nebula would have been cold enough to allow the volatile substances such as water, ammonia and methane to freeze into

THE PLANET JUPITER, seen here from the spacecraft *Pioneer 10*, is composed of much the same elements in the same proportion as the sun. It is a representative of one of the three compositional classes of objects in the solar system. This class includes the sun and Saturn. The other two classes are the rocky objects such as the terrestrial (earthlike) planets and the objects composed of both rocky and icy material, such as the nuclei of comets. In this picture the cloud belts of Jupiter are resolved in a wealth of detail invisible from the earth. The planet appears gibbous because of the position of the spacecraft with respect to the planet and to the sun, a view that can never be seen from the earth. The terminator, the line between the sunlit hemisphere and the dark one, is at the left of the image. The picture was made at a distance of 2,020,000 kilometers from the planet at 20:58 Universal Time on December 2, 1973, about 30 hours before the spacecraft's closest approach of 130,000 kilometers. It was made with the imaging photopolarimeter experiment of the University of Arizona. The photopolarimeter consists of a telescope with an aperture of 2.5 centimeters (one inch) coupled to an optical system that splits the light from the planet into a blue channel and a red one. The device scanned the planet as the spacecraft spun on its axis 4.8 times per minute, building up the image in a raster pattern somewhat like the image on a television set. The photopolarimeter alternated between the blue channel and the red channel every half millisecond and telemetered the raw data for each channel to the earth. At the University of Arizona the image from each channel was rectified to reduce distortion and increase its sharpness; then the two images were combined to produce a color composite.

OBJECT	DISTANCE FROM SUN (MILLIONS OF MILES)	DIAMETER (THOUSANDS OF MILES)	MASS (EARTH = 1)
SUN	–	867	343,000
MERCURY	36	3.0	.1
VENUS	67	7.6	.8
EARTH	93	7.9	1.0
MARS	142	4.2	.1
JUPITER	486	89	317.8
SATURN	892	76	95.2
URANUS	1,790	30	14.5
NEPTUNE	2,810	28	17.2
PLUTO	3,780	3.7	<.2?

NUMBER OF MOONS	DENSITY (WATER = 1)	ROTATION PERIOD (DAYS)	REVOLUTION AROUND SUN (YEARS)	AVERAGE TEMPERATURE (DEGREES K.)	GRAVITY (EARTH = 1)
–	1.4	27	–	5,800	28
0	5.4	55	.24	~ 600	.37
0	5.1	−243	.62	750	.89
1	5.5	1	1.00	180	1.00
2	3.9	1	1.88	140	.38
12	1.3	.4	11.86	128	2.65
10	.7	.4	29.48	105	1.14
5	1.6	.4	84.01	70	.96
2	2.3	.6	164.79	55	1.53
0	?	6.4	248.4	?	?

their ices, thus forming solid particles.

It is believed the sun formed when the center of the heated solar nebula became unstable and collapsed under the influence of its own gravity. The massive Jovian planets could have been formed by the same process. Alternatively they could have begun with a rocky and icy core so large that its gravitational attraction captured large masses of undifferentiated gaseous material. On this basis the similarity of their composition to the composition of the sun is easily understood.

The composition of the terrestrial planets and similar objects varies markedly with distance from the sun. There is a simple conceptual model, the equilibrium-condensation hypothesis, that can account for these differences. Let us assume that the chemical equilibrium between the dust and the gas in the solar nebula governed the composition of the solid materials that eventually accumulated into the planets, their satellites, the asteroids and the comets. On that basis one can calculate how the chemical composition of the condensed matter depends on its temperature at its formation. For this purpose one additional fact is needed: the relative abundances of the various chemical elements of which the primordial nebula was composed.

Our knowledge of the abundances of most elements in the galaxy is derived almost exclusively from the study of bodies in the solar system. The sun, however, is so much like most other stars that the relative amounts of the various elements in the solar system can be assumed to be typical of the large majority of stellar systems. Observations of the sun reveal that 99.996 percent of the mass of any sample of its material would consist of 15 elements. The least abundant element of the 15 (nickel) would still be commoner than all the other 80-odd elements put together.

There is a second model of how the terrestrial planets formed out of the solar nebula. The equilibrium-conden-

DATA ON THE PLANETS, incorporating information from various spacecraft missions as well as from ground-based observations, are displayed next to drawings of the planets that illustrate their relative sizes. Minus sign in front of rotation period of Venus indicates that the planet rotates in a direction opposite to the direction in which all other planets rotate. Tenth moon of Saturn, named Janus, was discovered in 1966.

DE-GREES KELVIN	EQUILIBRIUM-CONDENSATION MODEL	INHOMOGENEOUS-ACCRETION MODEL
1,600	1. Condensation of refractory oxides such as calcium oxide (CaO) and aluminum oxide (Al_2O_3) and also of titanium oxide and the rare-earth oxides	1. Condensation of refractory oxides such as calcium oxide (CaO) and aluminum oxide (Al_2O_3) and also of titanium oxide and the rare-earth oxides
1,300	2. Condensation of metallic iron-nickel alloy	2. Condensation of metallic iron-nickel alloy
1,200	3. Condensation of the mineral enstatite ($MgSiO_3$)	3. Condensation of the mineral enstatite ($MgSiO_3$)
1,000	4. Reaction of sodium (Na) with aluminum oxide and silicates to make feldspar and related minerals, and the deposition of potassium and the other alkali metals	4. Condensation of sodium oxide (Na_2O) and the other alkali-metal oxides at about 800 degrees K.
680	5. Reaction of hydrogen sulfide gas (H_2S) with metallic iron to make the sulfide mineral troilite (FeS)	
1,200 –490	6. Progressive oxidation of the remaining metallic iron to ferrous oxide (FeO), which in turn reacts with enstatite to make olivine (Fe_2SiO_4 and Mg_2SiO_4)	
550	7 Combination of water vapor (H_2O) with calcium-bearing minerals to make tremolite	
425	8. Combination of water vapor with olivine to make serpentine	
175	9. Condensation of water ice	5. Condensation of water ice (H_2O)
150	10. Reaction of ammonia gas (NH_3) with water ice to make the solid hydrate $NH_3 \cdot H_2O$	6. Condensation of ammonium hydrosulfide (NH_4SH)
120	11 Partial reaction of methane gas (CH_4) with water ice to make the solid hydrate CH_4 $7H_2O$	7 Condensation of ammonia ice (NH_3)
65	12. Condensation of argon (Ar) and leftover methane gas into solid argon and methane	8. Condensation of solid argon (Ar) and methane
20	13. Condensation of neon (Ne) and hydrogen, leading to 75-percent-complete condensation of solar materials	9. Condensation of neon (Ne) and hydrogen, leading to 75-percent-complete condensation of solar materials
~1	14. Condensation of helium (He) into liquid	10. Condensation of helium (He) into liquid

MAJOR REACTIONS that would have occurred in the formation of the solar system according to two different hypotheses are shown with respect to temperature in the primordial solar nebula. The equilibrium-condensation model assumes that the nebula cooled very slowly as the planets were forming. The inhomogeneous-accretion model assumes just the opposite: that the nebula cooled quickly with respect to the rate at which the planets were forming. The two assumptions result in radically different predictions for the composition of the planets and satellites. In the author's view the equilibrium-condensation model seems to describe the planets of the solar system more accurately. Color area indicates two steps of each hypothesis that probably did not occur because the temperature never fell that low.

sation model assumes that there were no substantial changes in the temperature of the nebula at the location of planets that were accreting solid material onto their surface. Perhaps the material accreted rapidly in comparison with the rate at which the nebula was cooling. Or perhaps the composition of the solid material was determined by the temperature when it had last reacted with the gases of the nebula; after that time the gases could have dissipated and the solids could have accreted slowly onto the planets, each of which would have been homogeneous in composition. In either case the chemical equilibrium between the gas and the dust would determine the composition of both the gases and the condensates present at any time.

Comparison of the Two Models

The second model, the inhomogeneous-accretion hypothesis, is exactly the opposite of the first. Here the solar nebula cooled rapidly in comparison with the rate at which the planets accreted solid material. Since the composition of the material would have been determined by its temperature its accretion onto a planet could have given rise to onionlike layers of different condensates. It is important to add that the two hypotheses are extreme cases; the truth may lie anywhere in between them. Since we have no a priori grounds for choosing between them, however, we should understand the consequences of both extremes.

The results of the two assumptions are quite different [see illustration at left]. If one starts with a temperature of 2,000 degrees K. in the solar nebula, the equilibrium-condensation hypothesis would lead first to the condensation of refractory compounds containing calcium oxide (CaO), aluminum oxide (Al_2O_3), rare-earth oxides and so on. Then when the nebula had cooled to about 1,500 degrees, a metallic iron-nickel alloy similar to the one found in meteorites would condense. That step would be followed by the condensation of enstatite ($MgSiO_3$) and then by the formation of various minerals such as feldspar through the deposition of sodium, potassium and the other alkali metals. As the temperature dropped to 680 degrees metallic iron would be corroded by hydrogen sulfide gas (H_2S) to make the mineral troilite (FeS); the remaining iron would be progressively oxidized to form minerals such as olivine, which would combine with water vapor at a still lower temperature to yield serpentine. Finally, as the temperature of the nebula dropped below 170 degrees, water vapor would condense as ice. The ice would later react with ammonia gas to make a hydrate, written $NH_3 \cdot H_2O$. At 100 degrees water ice would also combine with some of the methane gas in the nebula to yield another hydrate: $CH_4 \cdot 7H_2O$. Eventually argon and the leftover methane gas would freeze out at about 60 degrees. If the temperature continued to drop to as low as 10 degrees, neon and hydrogen would condense. Finally even helium would condense if the temperature ever fell below one degree. There is, however, no evidence that the temperature within the solar nebula was ever that low.

How does this sequence of events differ from the sequence predicted by the inhomogeneous-accretion hypothesis? One important difference is that those minerals that can be formed only by chemical reactions between gases and previously formed minerals cannot be made. This constraint rules out processes such as the corrosion of metallic iron by water vapor and by hydrogen sulfide gas, and the reaction of water ice with ammonia and methane to form the hydrates $NH_3 \cdot H_2O$ and $CH_4 \cdot 7H_2O$. Thus the sequence of chemical reactions in the process of inhomogeneous accretion is far simpler than the one in the process of equilibrium condensation.

At first the condensation of the refractory oxides, of the iron-nickel alloy and of enstatite would proceed in the same way as they would in the equilibrium-condensation process. Thereafter sodium oxide (Na_2O) would condense. Then water, ammonium hydrosulfide (NH_4SH) and ammonia would freeze into their respective ices. At the end methane and argon would freeze. Again the temperature would probably never drop low enough to allow the condensation of neon, hydrogen and helium. It is clear, however, that the dependence of composition and density on temperature for the material formed by inhomogeneous accretion would be quite different from that for material formed by equilibrium condensation.

There are ways to check the predictions of both models against reality. For example, we have quite a lot of information about the composition and the structure of one planet: the earth. Which of the two hypotheses is better able to account for what is known about that planet? Since the two models predict the content of volatile substances in

the material that formed the planets, which hypothesis is in better accord with our knowledge of planetary atmospheres? Can we discriminate between the two hypotheses on the basis of the substantial amount of information we have about carbonaceous chondrites, the oldest and most primitive meteorites?

Evidence from Meteorites and Planets

Let us look first at the evidence from the carbonaceous chondrites. That evidence strongly favors the process of equilibrium condensation. Unlike the crust of the earth, chondrites are composed of unheated, unmelted, undifferentiated primordial material. The mineralogy and relative abundances of the elements of this preplanetary stuff are available for direct inspection. A mineral that is nearly universal in meteorites is troilite, or FeS. Troilite is an important feature of the sequence of events in equilibrium condensation, but it is absent from the sequence in inhomogeneous accretion. The next most widely distributed minerals are pyroxene and olivine, most often containing between 5 and 20 percent ferrous oxide (FeO). Ferrous oxide is an essential part of the equilibrium-condensation sequence.

Carbonaceous chondrites are rich in volatile materials, including a large quantity of water bound to other substances such as the mineral serpentine. It is an interesting fact that the spectra of many asteroids are indistinguishable from laboratory spectra of carbonaceous chondrites. That would be expected on the basis of the equilibrium-condensation model. Finally, carbonaceous chondrites are highly homogeneous. There is absolutely no tendency for the material in these meteorites to be strongly sorted into many different pure minerals; instead they tend to be remarkably well mixed.

The available data on the atmospheres of the planets are also much more easily understood on the basis of the equilibrium-condensation process than on that of the inhomogeneous-accretion one. The content of volatile elements in the planets formed by inhomogeneous accretion would be zero until the time when water had condensed out of the solar nebula; thereafter the condensed material would be more than 60 percent water! On the other hand, the equilibrium-condensation model predicts the right amount of water for the earth while leaving Venus extremely

arid; it also predicts a water content for Mars that is six times higher than that for the earth.

The amount of carbon in the planets is predicted well by both models. The source of the oxygen to make carbon dioxide on the terrestrial planets would apparently have to be ferrous oxide. The content of ferrous oxide in Venus, the earth and Mars predicted by the equilibrium-condensation model is sufficient in all three cases to make the observed amounts of carbon dioxide. Ferrous oxide, however, is totally absent in the sequence predicted by the inhomogeneous-accretion model.

The basic chemical and physical properties of the earth all favor the equilibrium-condensation model. The chemical composition of the earth as a whole cannot be distinguished from that of the iron-rich carbonaceous chondrites. Unfortunately for students of such problems, however, some 99 percent of the mass of the earth consists not of crustal material but of mantle material and core material, neither of which can be analyzed directly. One must be content with inferring the composition of these materials from remote observations of such properties as density and the velocity of seismic waves.

It is generally agreed that the earth's mantle largely consists of compounds containing magnesium oxide (MgO), silicon dioxide (SiO_2) and ferrous oxide. Currently it seems probable that the entire mantle is about 10 percent ferrous oxide. The mantle is divided into two distinct layers: the upper layer is mainly silicates of iron and magnesium and the lower one, where the pressure is higher, is largely a mixture of oxides of iron and magnesium.

The core is also divided into two layers. The inner core, which accounts for only 1 percent of the earth's mass, has a density that corresponds to the density of a solid iron-nickel alloy under immense pressure. The outer core is liquid and has a density appreciably less than that of liquid iron alone at the same pressure, indicating that there must be a lighter element mixed in. The lighter component has generally been thought to be either silicon or sulfur. It is a remarkable coincidence that the equilibrium-condensation model alone provides the correct abundance of sulfur to explain the observed density of the outer core. Moreover, the equilibrium-condensation model explains why the earth has a slightly higher density than Venus by predicting that the earth has sulfur in its core and Venus does not.

Whether or not Venus actually is deficient in sulfur compared with the earth is not yet known, but it is certainly true that Venus' atmosphere lacks detectable sulfur compounds. If Venus emits volcanic gases as the earth does, it should have in its atmosphere a readily detectable amount of carbonyl sulfide (COS). Carbonyl sulfide has not been found, however, even by very sensitive spectroscopic methods.

Internal Structure of the Planets

On the basis of all the evidence from meteorites and planets let us accept the equilibrium-condensation model as yielding a good approximation of the primordial diversity in the composition of the solar system. Now let us briefly survey the nature of the physical and chemical processes going on inside the planets and satellites in order to gauge what types of internal structure could have been formed in their parent bodies.

The easiest question to ask is how the present internal structure of differentiated objects such as the earth depends on the temperature at which they formed. (Small bodies such as meteoroids and asteroids that were never warm enough to differentiate would of course remain as homogeneous mixtures of their constituent minerals.) In order to answer this question we must retrace the steps of the equilibrium-condensation process as the temperature within the solar nebula dropped from 2,000 degrees K.

The first minerals formed, the refractory oxide minerals, would have given rise to a class of protoplanets with a high concentration of calcium, aluminum, titanium and the rare-earth elements. The protoplanets would also have been rich in uranium and thorium, and hence they would have been heated rapidly by the radioactive decay of those elements in their interior. Thus in spite of the fact that the refractory oxide minerals have high melting points, the temperature inside the protoplanets would have been high enough for them to melt readily. The present-day interior of a large body of this type would be mostly homogeneous. Sulfides, iron and iron oxides would be totally absent. It has recently been suggested by Don L. Anderson of the California Institute of Technology that the moon is a member of this com-

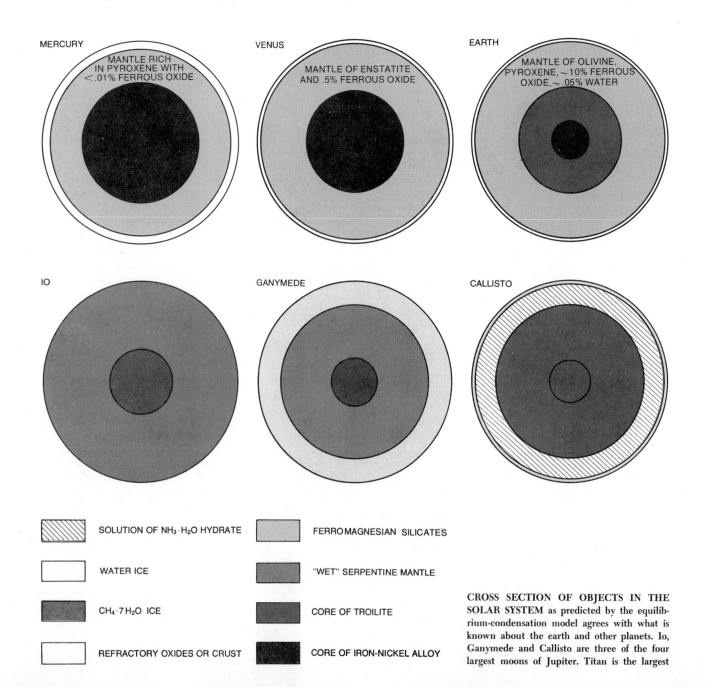

CROSS SECTION OF OBJECTS IN THE SOLAR SYSTEM as predicted by the equilibrium-condensation model agrees with what is known about the earth and other planets. Io, Ganymede and Callisto are three of the four largest moons of Jupiter. Titan is the largest

positional class. The moon would have had to have been extensively contaminated with sulfides and iron oxides, however, in order to account for its observed composition.

The second compositional class would have consisted of protoplanets formed just after the metallic iron-nickel alloy condensed out of the solar nebula. After these bodies had differentiated they would have had a thin crust of oxide minerals resting directly on a massive metallic core. They are not represented in the solar system today.

The protoplanets of the third class would have been formed after the temperature in the nebula had dropped far enough for enstatite to condense. The chemical makeup of a planet that came into being at this point would have been

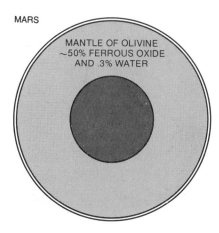

MARS

MANTLE OF OLIVINE
~50% FERROUS OXIDE
AND .3% WATER

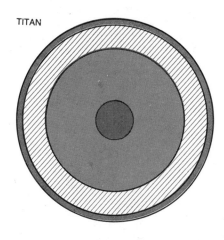

TITAN

moon of Saturn. "Wet" mantle of serpentine contains much water bound into the rock. The boundary between the inner mantle and the outer mantle of the earth is not shown because it is a change in the mineral structure and not a difference in chemical composition.

dominated by silicates. A crust of oxide minerals would have formed a thin layer on top of a mantle rich in pyroxene containing a negligible amount (less than .01 percent) of ferrous oxide. The planet Mercury is a member of this class.

The protoplanets of the fourth class would have differed from those in the third in that their crust would have retained alkali metals, resulting in a composition chemically similar to the crust of the earth except that the content of volatile elements would have been much lower. Such a body would have had a mantle containing perhaps .5 percent ferrous oxide. The bodies in this class would have corresponded to Venus.

The protoplanets of the fifth class would have retained sulfur as a constituent of troilite. On melting, the troilite would have sunk to form a dense outer core. Most of the mass of the core would have consisted of this sulfide melt rather than solid metal. The mantle, which would have been composed of olivine and pyroxene, would have contained at least 2 percent ferrous oxide. This class is not represented in the present solar system except perhaps in meteorites.

The sixth major class would have been made up of protoplanets that had formed when the temperature in the solar nebula had dropped enough (to about 600 degrees K.) for water to be retained, bound in the crystal structure of minerals such as tremolite. Up to .3 percent of the mass of these bodies could have been water. The earth, with a core rich in sulfur, its mantle containing some 10 percent ferrous oxide and its water content about .05 percent, belongs to this class.

Mars and Beyond

The seventh major class would have consisted of protoplanets that had formed after metallic iron had been completely oxidized. Such a planet would have contained some .3 percent water. Its olivine mantle would have been very rich in ferrous oxide, perhaps consisting of as much as 50 percent ferrous oxide. The core would have been either solid or liquid troilite, devoid of metallic iron. The resulting object would closely resemble Mars.

The protoplanets of the eighth class, which would have formed at a temperature low enough for the olivine in the mantle to have combined with the hydroxyl radical in the solar nebula to yield a "wet" mantle of serpentine, would have had a tremendously high water content: about 14 percent by mass. The

difference in the density between the minerals in the crust and those in the mantle would have essentially disappeared, but it is possible that a core of ferrous sulfide could have existed. A homogeneous parent body of this composition could have differentiated enough to form a core even at such a relatively low temperature as 400 degrees because the melting point of silicates with a large water content is fairly low. The asteroids probably belong to this eighth class. Calculations based on the known melting points of silicates should reveal whether or not such small objects could be warmed enough by the decay of radioactive elements in their interior for the minerals to be differentiated in their feeble gravitational field. If the asteroids are in fact like the carbonaceous chondrites, then perhaps the ordinary drier, less oxidized chondrites originated between the orbits of the earth and Mars instead of in the asteroid belt.

The protoplanets of the ninth class, formed below the condensation temperature of water, would have contained water ice. In such a body the water content, both as ice and as water bound in minerals, would have been about 60 percent by weight. Thus its density would have been quite low: only about 1.7 grams per cubic centimeter and less than a third the average density of the earth. Jupiter's largest satellite, Ganymede, may fall into this class.

The protoplanets of the 10th class would have retained ammonia as the solid hydrate $NH_3 \cdot H_2O$. A mixture of this hydrate with ice begins to melt at the low temperature of 173 degrees K., which is only a few degrees above the daytime surface temperature of Jupiter's second-largest satellite, Callisto. If a protoplanet in this class had ever been heated enough by radioactive decay to differentiate, it would have had a deep mantle of a liquid water-ammonia solution surmounted by a thin crust of water ice. During the differentiation a core of sulfides, oxides and hydrous (water-containing) silicates would have been formed as a sediment. The overall result would have been a stratified ball of mud and slush.

The protoplanets of the 11th class would have been about 4 percent methane by weight. The heat generated by radioactive decay would have melted the solid hydrate $CH_4 \cdot 7H_2O$, liberating methane gas to form an atmosphere. Differentiation would have given rise to a solid crust of the hydrate. A prime candidate for membership in this class is Saturn's largest satellite, Titan, which has a massive atmosphere rich in methane.

The protoplanets of the 12th and last class would have held solid methane as the temperature in the solar nebula dropped to almost 50 degrees K. Since carbon is nearly as abundant as oxygen in the solar system, and since the density of solid or liquid methane is very low (about .6 gram per cubic centimeter), the density of such a protoplanet would have been only one gram per c.c. Several of the smaller satellites of Saturn have been reported to have densities of near one gram per c.c., but the margin of error in such measurements is at least a factor of two. It is entirely possible that some of the satellites of Uranus or Neptune belong to this class, but their masses and radii will have to be determined by spacecraft flyby missions before we can confirm that speculation with any certainty.

I have mentioned the satellites of Jupiter, Saturn, Uranus and Neptune but not the planets themselves. There are reasons for the absence of the Jovian planets from the compositional classes. As we have seen, these classes are based on the temperature at which the protoplanetary material condensed out of the solar nebula according to the equilibrium-condensation model. Now, since the composition of Jupiter and Saturn is essentially the same as the composition of the sun, they could have formed just about anywhere in the solar nebula with no strong dependence on the temperature in that region. All that would be required is an initial protoplanetary core whose gravitational field was strong enough to accrete the undifferentiated solar material. Uranus, Neptune and Pluto are left off the list simply because not enough is known about their composition.

Important Results

One important result of the sequence of reactions I have been discussing is that the densities of the terrestrial planets, like their chemistry, can all be explained as a direct consequence of the variation of the temperature at which

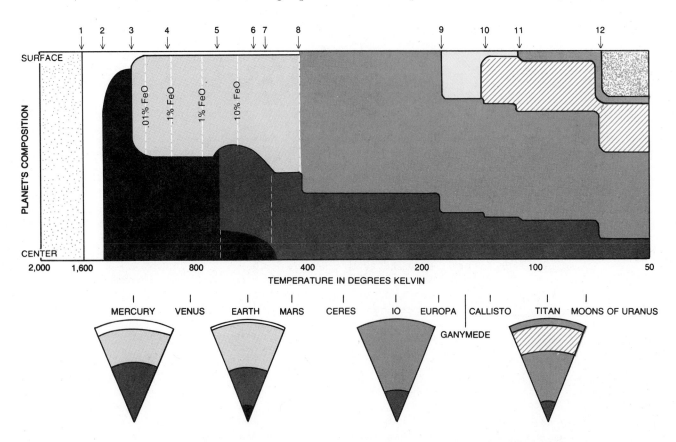

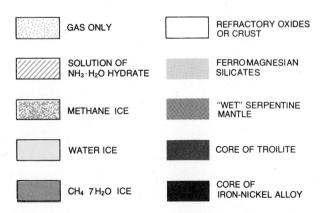

EQUILIBRIUM-CONDENSATION MODEL of the formation of the solar system is compared with the inhomogeneous-accretion model (*see illustration on the opposite page*). The left edge of the diagram, at 2,000 degrees Kelvin, represents the interior of the solar nebula; the right edge, at 50 degrees K., represents a distance of about 30 times the distance of the earth from the center of the nebula. Numbers across the top indicate the temperatures and the locations of the 12 major reactions of the equilibrium-condensation hypothesis. At bottom are shown the temperatures at which some of the objects in the solar system would have been formed. Vertical axis represents model's prediction of objects' internal composition from its center (*bottom*) to its surface (*top*) regardless of the specific distance in miles. Pie-shaped sections at bottom show the cross section of four objects in the solar system (Mercury, the earth, Io and Titan) that formed at different temperatures. FeO is ferrous oxide. Small area of iron-nickel alloy core represents solid inner core of the earth; small area of troilite is melted outer core.

they were formed according to their distance from the sun. It has often been argued that, by analogy with a hypothesis about the formation of meteorites, the different densities of the terrestrial planets are due to a physical process that fractionated the material in such a way that metals were preferentially accreted in one case (Mercury) and silicates were preferentially accreted in the others. Such preferential accretion, as it is seen in meteorites, might have been the result of differences in the magnetic or electrical properties of the material, or in how well the accreted matter adhered to the surface of the protoplanet. Since the sequence of reactions of the equilibrium-condensation model leads automatically to the high density of Mercury, we are led to deny that any fractionation process affected the planets. Planets are so massive that they must have accreted their material by gravitation, a totally nonselective process. Temperature alone determined the composition of the accreted material. On the other hand, the parent bodies of meteorites, which could have had a mass as small as 10^{-12} times the mass of a planet, would have had only a feeble gravitational attraction. They probably accreted their material almost entirely through selective interparticle forces such as magnetism. Thus the paradox vanishes.

In summary, the equilibrium-condensation model shows very simply how the density and the volatile content of the solid material in the solar system are related to the temperature at which the material was formed. The model leads to a large variety of quantitative predictions that can be tested in the laboratory, by astronomical observations and by planetary probes. Within a single chemical sequence we can now interrelate such seemingly diverse matters as the bulk composition of the solar nebula, the internal structure of the planets and the content of volatile compounds in the planets.

Perhaps the most striking result of this approach, however, is the underlying vision of an intimate interrelationship among all the objects in the solar

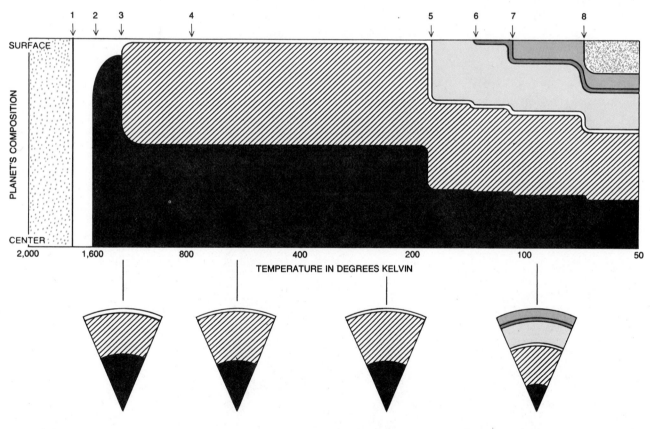

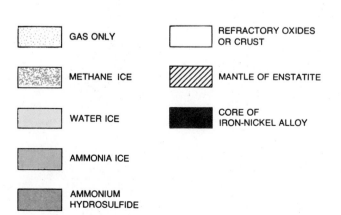

GAS ONLY

METHANE ICE

WATER ICE

AMMONIA ICE

AMMONIUM HYDROSULFIDE

REFRACTORY OXIDES OR CRUST

MANTLE OF ENSTATITE

CORE OF IRON-NICKEL ALLOY

INHOMOGENEOUS-ACCRETION MODEL of the formation of the solar system predicts quite different compositions of the planets as they formed at various temperatures. This diagram is read in the same way as the one on the opposite page. The eight major reactions of the inhomogeneous-accretion model are shown at the top. The pie-shaped sections at the bottom represent cross sections of hypothetical planets that would have been formed according to the inhomogeneous-accretion model at the same temperatures that Mercury, the earth, Io and Titan were formed at according to the equilibrium-condensation model. The hypothetical Mercury would have had the same structure. The earth, however, would have had a mantle of the mineral enstatite and a solid core of iron-nickel alloy, a prediction that is not supported by seismic-wave evidence. The hypothetical Io would have had the same composition as Venus, the earth and the asteroids even though all of them formed at radically different temperatures and distances from the sun. Titan would have had half a dozen onionlike layers of different materials.

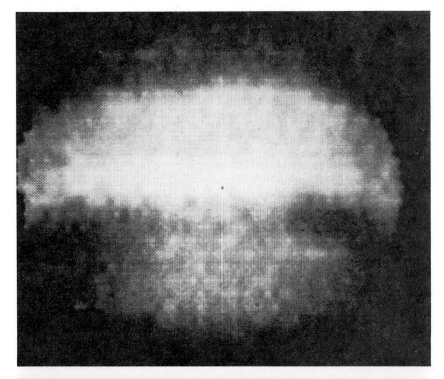

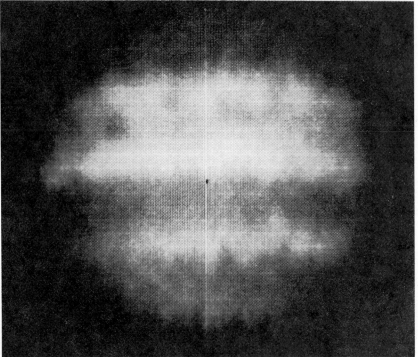

HEAT MAPS OF JUPITER at two different wavelengths were made from *Pioneer 10* with the infrared radiometer experiment of the California Institute of Technology, directed by Guido Munch. The radiometer is a reflecting telescope with an aperture of three inches mounted rigidly on the spacecraft. The spin of *Pioneer 10* and its motion in its trajectory past Jupiter allowed the radiometer to scan the planet and build up its images. The top picture was made at a wavelength of 20 microns and measured the heat being emitted at a distance of 40 kilometers below the top of Jupiter's atmosphere; the bottom picture was made at a wavelength of 40 microns and measured the heat being emitted at a distance of 50 kilometers below the top of the atmosphere. Both were made on December 3, 1973, during the spacecraft's closest approach to the planet. The warmest (lightest) areas are eight degrees K. warmer than the coolest (darkest) areas. The warmest areas correspond to the dark belts of Jupiter; coolest areas correspond to the bright belts. Planet is warmest near equator. It is estimated that the planet has an average temperature close to 128 degrees K. It radiates more than twice the amount of thermal energy it receives from the sun, confirming observations from the earth indicating that planet generates some heat of its own.

system. We see that there is strong evidence that the planets formed from a medium of uniform elemental composition. We see that the same qualitative processes of condensation and accretion are at work in the same general way throughout the entire system. The presence or absence of sulfur gases in the atmosphere of Venus, the density of the core of Mars, the electrical conductivity of the mantle of Callisto, the atmospheric pressure at the surface of Titan, the radius and mass of the satellites of Uranus—all tangibly influence our comprehension of the origin, composition and structure of our own planet. In this era in the history of geology we have seen how important it is to study the origins of all the chemical elements in order to understand the origin of any one of them, and we have been obliged to study global tectonics in order to understand the origin of any one mountain. Now we must be prepared to learn from comparative studies of the planets exactly what processes affect planets in general and how planets evolve. From this kind of investigation we can learn for the first time how the earth originated, evolved and attained its present structure.

Another long-range benefit of such comparative studies will be to provide insight into how frequently planets might be formed around other stars, and what the composition of those planets might be. Since the composition of most stars is indistinguishable from that of the sun, and since present theories of the origin of the solar system suggest that the formation of planets is a normal by-product of the formation of stars, it would not be surprising to find that planets similar to those in the solar system are common throughout our galaxy and other spiral galaxies. It is interesting to speculate whether or not planets belonging to some compositional classes that are not seen in the solar system actually exist in other planetary systems. One such example would be a planet made up of iron and refractory oxides that was far denser than Mercury; another would be a very earthlike planet that was devoid of water; a third would be a true planet composed of material rich in volatile compounds such as the material ascribed to the asteroids.

The next few years will provide numerous opportunities for directly testing these predictions. A vigorous program for exploring the planets is the only way to realize the full benefit of the opportunities. It is the logical next step in our efforts to understand the planet on which we live.

The Origin of Life

by George Wald
August 1954

How did living matter first arise on the earth? As natural scientists learn more about nature they are returning to a hypothesis their predecessors gave up almost a century ago: spontaneous generation

About a century ago the question, How did life begin?, which has interested men throughout their history, reached an impasse. Up to that time two answers had been offered: one that life had been created supernaturally, the other that it arises continually from the nonliving. The first explanation lay outside science; the second was now shown to be untenable. For a time scientists felt some discomfort in having no answer at all. Then they stopped asking the question.

Recently ways have been found again to consider the origin of life as a scientific problem—as an event within the order of nature. In part this is the result of new information. But a theory never rises of itself, however rich and secure the facts. It is an act of creation. Our present ideas in this realm were first brought together in a clear and defensible argument by the Russian biochemist A. I. Oparin in a book called *The Origin of Life*, published in 1936. Much can be added now to Oparin's discussion, yet it provides the foundation upon which all of us who are interested in this subject have built.

The attempt to understand how life originated raises a wide variety of scientific questions, which lead in many and diverse directions and should end by casting light into many obscure corners. At the center of the enterprise lies the hope not only of explaining a great past event—important as that should be—but of showing that the explanation is workable. If we can indeed come to understand how a living organism arises from the nonliving, we should be able to construct one—only of the simplest description, to be sure, but still recognizably alive. This is so remote a possibility now

that one scarcely dares to acknowledge it; but it is there nevertheless.

One answer to the problem of how life originated is that it was created. This is an understandable confusion of nature with technology. Men are used to making things; it is a ready thought that those things not made by men were made by a superhuman being. Most of the cultures we know contain mythical accounts of a supernatural creation of life. Our own tradition provides such an account in the opening chapters of *Genesis*. There we are told that beginning on the third day of the Creation, God brought forth living creatures—first plants, then fishes and birds, then land animals and finally man.

Spontaneous Generation

The more rational elements of society, however, tended to take a more naturalistic view of the matter. One had only to accept the evidence of one's senses to know that life arises regularly from the nonliving: worms from mud, maggots from decaying meat, mice from refuse of various kinds. This is the view that came to be called spontaneous generation. Few scientists doubted it. Aristotle, Newton, William Harvey, Descartes, van Helmont, all accepted spontaneous generation without serious question. Indeed, even the theologians—witness the English Jesuit John Turberville Needham—could subscribe to this view, for *Genesis* tells us, not that God created plants and most animals directly, but that He bade the earth and waters to bring them forth; since this directive was never rescinded, there is nothing heretical in believing that the process has continued.

But step by step, in a great controversy that spread over two centuries, this belief was whittled away until nothing remained of it. First the Italian Francesco Redi showed in the 17th century that meat placed under a screen, so that flies cannot lay their eggs on it, never develops maggots. Then in the following century the Italian abbé Lazzaro Spallanzani showed that a nutritive broth, sealed off from the air while boiling, never develops microorganisms, and hence never rots. Needham objected that by too much boiling Spallanzani had rendered the broth, and still more the air above it, incompatible with life. Spallanzani could defend his broth; when he broke the seal of his flasks, allowing new air to rush in, the broth promptly began to rot. He could find no way, however, to show that the air in the sealed flask had not been vitiated. This problem finally was solved by Louis Pasteur in 1860, with a simple modification of Spallanzani's experiment. Pasteur too used a flask containing boiling broth, but instead of sealing off the neck he drew it out in a long, S-shaped curve with its end open to the air. While molecules of air could pass back and forth freely, the heavier particles of dust, bacteria and molds in the atmosphere were trapped on the walls of the curved neck and only rarely reached the broth. In such a flask the broth seldom was contaminated; usually it remained clear and sterile indefinitely.

This was only one of Pasteur's experiments. It is no easy matter to deal with so deeply ingrained and common-sense a belief as that in spontaneous generation. One can ask for nothing better in such a pass than a noisy and stubborn opponent, and this Pasteur had in the naturalist Félix Pouchet, whose arguments before the French Academy of Sciences drove Pasteur to more and more rigorous experiments. When he had finished, nothing remained of the belief in spontaneous generation.

We tell this story to beginning students of biology as though it represents a triumph of reason over mysticism. In fact it is very nearly the opposite. The reasonable view was to believe in spontaneous generation; the only alternative, to believe in a single, primary act of supernatural creation. There is no third position. For this reason many scientists a century ago chose to regard the belief in spontaneous generation as a "philosophical necessity." It is a symptom of the philosophical poverty of our time that this necessity is no longer appreciated. Most modern biologists, having reviewed with satisfaction the downfall of the spontaneous generation hypothesis, yet unwilling to accept the alternative belief in special creation, are left with nothing.

I think a scientist has no choice but to approach the origin of life through a hypothesis of spontaneous generation. What the controversy reviewed above showed to be untenable is only the belief that living organisms arise spontaneously under present conditions. We have now to face a somewhat different problem: how organisms may have arisen spontaneously under different conditions in some former period, granted that they do so no longer.

The Task

To make an organism demands the right substances in the right proportions and in the right arrangement. We do not think that anything more is needed—but that is problem enough.

The substances are water, certain salts—as it happens, those found in the ocean—and carbon compounds. The latter are called *organic* compounds because they scarcely occur except as products of living organisms.

Organic compounds consist for the most part of four types of atoms: carbon, oxygen, nitrogen and hydrogen. These four atoms together constitute about 99 per cent of living material, for hydrogen and oxygen also form water. The organic compounds found in organisms fall mainly into four great classes: carbohydrates, fats, proteins and nucleic acids. The illustrations on this and the next three pages give some notion of their composition and degrees of complexity. The fats are simplest, each consisting of three fatty acids joined to glycerol. The starches and glycogens are made of sugar units strung together to form long straight and branched chains. In general only one type of sugar appears in a single starch or glycogen; these molecules are large, but still relatively simple. The principal function of carbohydrates and fats in the organism is to serve as fuel—as a source of energy.

The nucleic acids introduce a further level of complexity. They are very large structures, composed of aggregates of at least four types of unit—the nucleotides—brought together in a great variety of proportions and sequences. An almost endless variety of different nucleic acids is possible, and specific differences among them are believed to be of the highest importance. Indeed, these structures are thought by many to be the main constituents of the genes, the bearers of hereditary constitution.

Variety and specificity, however, are most characteristic of the proteins, which include the largest and most complex molecules known. The units of

which their structure is built are about 25 different amino acids. These are strung together in chains hundreds to thousands of units long, in different proportions, in all types of sequence, and with the greatest variety of branching and folding. A virtually infinite number of different proteins is possible. Organisms seem to exploit this potentiality, for no two species of living organism, animal or plant, possess the same proteins.

Organic molecules therefore form a large and formidable array, endless in variety and of the most bewildering complexity. One cannot think of having organisms without them. This is precisely the trouble, for to understand how organisms originated we must first of all explain how such complicated molecules could come into being. And that is only the beginning. To make an organism requires not only a tremendous variety of these substances, in adequate amounts and proper proportions, but also just the right arrangement of them. Structure here is as important as composition—and what a complication of structure! The most complex machine man has devised—say an electronic brain—is child's play compared with the simplest of living organisms. The especially trying thing is that complexity here involves such small dimensions. It is on the molecular level; it consists of a detailed fitting of molecule to molecule such as no chemist can attempt.

The Possible and Impossible

One has only to contemplate the magnitude of this task to concede that the spontaneous generation of a living organism is impossible. Yet here we are—as a result, I believe, of spontaneous generation. It will help to digress for a mo-

CARBOHYDRATES comprise one of the four principal kinds of carbon compound found in living matter. This structural formula represents part of a characteristic carbohydrate. It is a polysaccharide consisting of six-carbon sugar units, three of which are shown.

ment to ask what one means by "impossible."

With every event one can associate a probability—the chance that it will occur. This is always a fraction, the proportion of times the event occurs in a large number of trials. Sometimes the probability is apparent even without trial. A coin has two faces; the probability of tossing a head is therefore 1/2. A die has six faces; the probability of throwing a deuce is 1/6. When one has no means of estimating the probability beforehand, it must be determined by counting the fraction of successes in a large number of trials.

Our everyday concept of what is impossible, possible or certain derives from our experience: the number of trials that may be encompassed within the space of a human lifetime, or at most within recorded human history. In this colloquial, practical sense I concede the spontaneous origin of life to be "impossible." It is impossible as we judge events in the scale of human experience.

We shall see that this is not a very meaningful concession. For one thing, the time with which our problem is concerned is geological time, and the whole extent of human history is trivial in the balance. We shall have more to say of this later.

But even within the bounds of our own time there is a serious flaw in our judgment of what is possible. It sounds impressive to say that an event has never been observed in the whole of human history. We should tend to regard such an event as at least "practically" impossible, whatever probability is assigned to it on abstract grounds. When we look a little further into such a statement, however, it proves to be almost meaningless. For men are apt to reject reports of very improbable occurrences. Persons of good

judgment think it safer to distrust the alleged observer of such an event than to believe him. The result is that events which are merely very extraordinary acquire the reputation of never having occurred at all. Thus the highly improbable is made to appear impossible.

To give an example: Every physicist knows that there is a very small probability, which is easily computed, that the table upon which I am writing will suddenly and spontaneously rise into the air. The event requires no more than that the molecules of which the table is composed, ordinarily in random motion in all directions, should happen by chance to move in the same direction. Every physicist concedes this possibility; but try telling one that you have seen it happen. Recently I asked a friend, a Nobel laureate in physics, what he would say if I told him that. He laughed and said that he would regard it as more probable that I was mistaken than that the event had actually occurred.

We see therefore that it does not mean much to say that a very improbable event has never been observed. There is a conspiracy to suppress such observations, not among scientists alone, but among all judicious persons, who have learned to be skeptical even of what they see, let alone of what they are told. If one group is more skeptical than others, it is perhaps lawyers, who have the harshest experience of the unreliability of human evidence. Least skeptical of all are the scientists, who, cautious as they are, know very well what strange things are possible.

A final aspect of our problem is very important. When we consider the spontaneous origin of a living organism, this is not an event that need happen again and again. It is perhaps enough for it to happen once. The probability with

which we are concerned is of a special kind; it is the probability that an event occur *at least once*. To this type of probability a fundamentally important thing happens as one increases the number of trials. However improbable the event in a single trial, it becomes increasingly probable as the trials are multiplied. Eventually the event becomes virtually inevitable. For instance, the chance that a coin will not fall head up in a single toss is 1/2. The chance that no head will appear in a series of tosses is $1/2 \times 1/2 \times 1/2 \ldots$ as many times over as the number of tosses. In 10 tosses the chance that no head will appear is therefore 1/2 multiplied by itself 10 times, or 1/1,000. Consequently the chance that a head will appear at least once in 10 tosses is 999/1,000. Ten trials have converted what started as a modest probability to a near certainty.

The same effect can be achieved with any probability, however small, by multiplying sufficiently the number of trials. Consider a reasonably improbable event, the chance of which is 1/1,000. The chance that this will not occur in one trial is 999/1,000. The chance that it won't occur in 1,000 trials is 999/1,000 multiplied together 1,000 times. This fraction comes out to be 37/100. The chance that it will happen at least once in 1,000 trials is therefore one minus this number—63/100—a little better than three chances out of five. One thousand trials have transformed this from a highly improbable to a highly probable event. In 10,000 trials the chance that this event will occur at least once comes out to be 19,999/20,000. It is now almost inevitable.

It makes no important change in the argument if we assess the probability that an event occur at least two, three, four or some other small number of

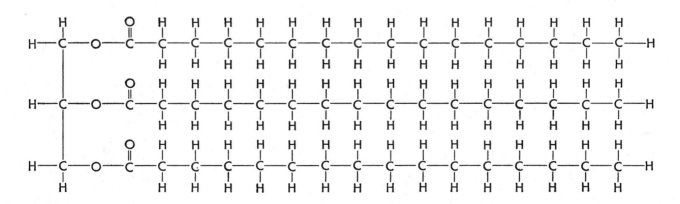

FATS are a second kind of carbon compound found in living matter. This formula represents the whole molecule of palmitin, one of the commonest fats. The molecule consists of glycerol (*11 atoms at the far left*) and fatty acids (*hydrocarbon chains at the right*).

times rather than at least once. It simply means that more trials are needed to achieve any degree of certainty we wish. Otherwise everything is the same.

In such a problem as the spontaneous origin of life we have no way of assessing probabilities beforehand, or even of deciding what we mean by a trial. The origin of a living organism is undoubtedly a stepwise phenomenon, each step with its own probability and its own conditions of trial. Of one thing we can be sure, however: whatever constitutes a trial, more such trials occur the longer the interval of time.

The important point is that since the origin of life belongs in the category of at-least-once phenomena, time is on its side. However improbable we regard this event, or any of the steps which it involves, given enough time it will almost certainly happen at least once. And for life as we know it, with its capacity for growth and reproduction, once may be enough.

Time is in fact the hero of the plot. The time with which we have to deal is of the order of two billion years. What we regard as impossible on the basis of human experience is meaningless here. Given so much time, the "impossible" becomes possible, the possible probable, and the probable virtually certain. One has only to wait: time itself performs the miracles.

Organic Molecules

This brings the argument back to its first stage: the origin of organic compounds. Until a century and a quarter ago the only known source of these substances was the stuff of living organisms. Students of chemistry are usually told that when, in 1828, Friedrich Wöhler synthesized the first organic compound, urea, he proved that organic compounds do not require living organisms to make

them. Of course it showed nothing of the kind. Organic chemists are alive; Wöhler merely showed that they can make organic compounds externally as well as internally. It is still true that with almost negligible exceptions all the organic matter we know is the product of living organisms.

The almost negligible exceptions, however, are very important for our argument. It is now recognized that a constant, slow production of organic molecules occurs without the agency of living things. Certain geological phenomena yield simple organic compounds. So, for example, volcanic eruptions bring metal carbides to the surface of the earth, where they react with water vapor to yield simple compounds of carbon and hydrogen. The familiar type of such a reaction is the process used in old-style bicycle lamps in which acetylene is made by mixing iron carbide with water.

Recently Harold Urey, Nobel laureate in chemistry, has become interested in the degree to which electrical discharges in the upper atmosphere may promote the formation of organic compounds. One of his students, S. L. Miller, performed the simple experiment of circulating a mixture of water vapor, methane (CH_4), ammonia (NH_3) and hydrogen—all gases believed to have been present in the early atmosphere of the earth—continuously for a week over an electric spark. The circulation was maintained by boiling the water in one limb of the apparatus and condensing it in the other. At the end of the week the water was analyzed by the delicate method of paper chromatography. It was found to have acquired a mixture of amino acids! Glycine and alanine, the simplest amino acids and the most prevalent in proteins, were definitely identified in the solution, and there were indications it contained aspartic acid and two others. The yield was surprisingly

high. This amazing result changes at a stroke our ideas of the probability of the spontaneous formation of amino acids.

A final consideration, however, seems to me more important than all the special processes to which one might appeal for organic syntheses in inanimate nature.

It has already been said that to have organic molecules one ordinarily needs organisms. The synthesis of organic substances, like almost everything else that happens in organisms, is governed by the special class of proteins called enzymes—the organic catalysts which greatly accelerate chemical reactions in the body. Since an enzyme is not used up but is returned at the end of the process, a small amount of enzyme can promote an enormous transformation of material.

Enzymes play such a dominant role in the chemistry of life that it is exceedingly difficult to imagine the synthesis of living material without their help. This poses a dilemma, for enzymes themselves are proteins, and hence among the most complex organic components of the cell. One is asking, in effect, for an apparatus which is the unique property of cells in order to form the first cell.

This is not, however, an insuperable difficulty. An enzyme, after all, is only a catalyst; it can do no more than change the *rate* of a chemical reaction. It cannot make anything happen that would not have happened, though more slowly, in its absence. Every process that is catalyzed by an enzyme, and every product of such a process, would occur without the enzyme. The only difference is one of rate.

Once again the essence of the argument is time. What takes only a few moments in the presence of an enzyme or other catalyst may take days, months or years in its absence; but given time, the end result is the same.

NUCLEIC ACIDS are a third kind of carbon compound. This is part of desoxyribonucleic acid, the backbone of which is five-carbon sugars alternating with phosphoric acid. The letter R is any one of four nitrogenous bases, two purines and two pyrimidines.

Indeed, this great difficulty in conceiving of the spontaneous generation of organic compounds has its positive side. In a sense, organisms demonstrate to us what organic reactions and products are *possible*. We can be certain that, given time, all these things must occur. Every substance that has ever been found in an organism displays thereby the finite probability of its occurrence. Hence, given time, it should arise spontaneously. One has only to wait.

It will be objected at once that this is just what one cannot do. Everyone knows that these substances are highly perishable. Granted that, within long spaces of time, now a sugar molecule, now a fat, now even a protein might form spontaneously, each of these molecules should have only a transitory existence. How are they ever to accumulate; and, unless they do so, how form an organism?

We must turn the question around. What, in our experience, is known to destroy organic compounds? Primarily two agencies: decay and the attack of oxygen. But decay is the work of living organisms, and we are talking of a time before life existed. As for oxygen, this introduces a further and fundamental section of our argument.

It is generally conceded at present that the early atmosphere of our planet contained virtually no free oxygen. Almost all the earth's oxygen was bound in the form of water and metal oxides. If this were not so, it would be very difficult to imagine how organic matter could accumulate over the long stretches of time that alone might make possible the spontaneous origin of life. This is a crucial point, therefore, and the statement that the early atmosphere of the planet was virtually oxygen-free comes forward so opportunely as to raise a suspicion of special pleading. I have for this reason taken care to consult a number of

geologists and astronomers on this point, and am relieved to find that it is well defended. I gather that there is a widespread though not universal consensus that this condition did exist. Apparently something similar was true also for another common component of our atmosphere—carbon dioxide. It is believed that most of the carbon on the earth during its early geological history existed as the element or in metal carbides and hydrocarbons; very little was combined with oxygen.

This situation is not without its irony. We tend usually to think that the environment plays the tune to which the organism must dance. The environment is given; the organism's problem is to adapt to it or die. It has become apparent lately, however, that some of the most important features of the physical environment are themselves the work of living organisms. Two such features have just been named. The atmosphere of our planet seems to have contained no oxygen until organisms placed it there by the process of plant photosynthesis. It is estimated that at present all the oxygen of our atmosphere is renewed by photosynthesis once in every 2,000 years, and that all the carbon dioxide passes through the process of photosynthesis once in every 300 years. In the scale of geological time, these intervals are very small indeed. We are left with the realization that all the oxygen and carbon dioxide of our planet are the products of living organisms, and have passed through living organisms over and over again.

Forces of Dissolution

In the early history of our planet, when there were no organisms or any free oxygen, organic compounds should have been stable over very long periods. This is the crucial difference between

the period before life existed and our own. If one were to specify a single reason why the spontaneous generation of living organisms was possible once and is so no longer, this is the reason.

We must still reckon, however, with another destructive force which is disposed of less easily. This can be called spontaneous dissolution—the counterpart of spontaneous generation. We have noted that any process catalyzed by an enzyme can occur in time without the enzyme. The trouble is that the processes which synthesize an organic substance are reversible: any chemical reaction which an enzyme may catalyze will go backward as well as forward. We have spoken as though one has only to wait to achieve syntheses of all kinds; it is truer to say that what one achieves by waiting is *equilibria* of all kinds—equilibria in which the synthesis and dissolution of substances come into balance.

In the vast majority of the processes in which we are interested the point of equilibrium lies far over toward the side of dissolution. That is to say, spontaneous dissolution is much more probable, and hence proceeds much more rapidly, than spontaneous synthesis. For example, the spontaneous union, step by step, of amino acid units to form a protein has a certain small probability, and hence might occur over a long stretch of time. But the dissolution of the protein or of an intermediate product into its component amino acids is much more probable, and hence will go ever so much more rapidly. The situation we must face is that of patient Penelope waiting for Odysseus, yet much worse: each night she undid the weaving of the preceding day, but here a night could readily undo the work of a year or a century.

How do present-day organisms manage to synthesize organic compounds against the forces of dissolution? They do so by a continuous expenditure of

PROTEINS are a fourth kind of carbon compound found in living matter. This formula represents part of a polypeptide chain, the backbone of a protein molecule. The chain is made up of amino acids. Here the letter R represents the side chains of these acids.

FILAMENTS OF COLLAGEN, a protein which is usually found in long fibrils, were dispersed by placing them in dilute acetic acid. This electron micrograph, which enlarges the filaments 75,000 times, was made by Jerome Gross of the Harvard Medical School.

energy. Indeed, living organisms commonly do better than oppose the forces of dissolution; they grow in spite of them. They do so, however, only at enormous expense to their surroundings. They need a constant supply of material and energy merely to maintain themselves, and much more of both to grow and reproduce. A living organism is an intricate machine for performing exactly this function. When, for want of fuel or through some internal failure in its mechanism, an organism stops actively synthesizing itself in opposition to the processes which continuously decompose it, it dies and rapidly disintegrates.

What we ask here is to synthesize organic molecules without such a machine. I believe this to be the most stubborn problem that confronts us—the weakest link at present in our argument. I do not think it by any means disastrous, but it calls for phenomena and forces some of which are as yet only partly understood and some probably still to be discovered.

Forces of Integration

At present we can make only a beginning with this problem. We know that it is possible on occasion to protect molecules from dissolution by precipitation or by attachment to other molecules. A wide variety of such precipitation and "trapping" reactions is used in modern chemistry and biochemistry to promote syntheses. Some molecules appear to acquire a degree of resistance to disintegration simply through their size. So, for example, the larger molecules composed of amino acids—polypeptides and proteins—seem to display much less tendency to disintegrate into their units than do smaller compounds of two or three amino acids.

Again, many organic molecules dis-

play still another type of integrating force—a spontaneous impulse toward structure formation. Certain types of fatty molecules—lecithins and cephalins—spin themselves out in water to form highly oriented and well-shaped structures—the so-called myelin figures. Proteins sometimes orient even in solution, and also may aggregate in the solid state in highly organized formations. Such spontaneous architectonic tendencies are still largely unexplored, particularly as they may occur in complex mixtures of substances, and they involve forces the strength of which has not yet been estimated.

What we are saying is that possibilities exist for opposing *intra*molecular dissolution by *inter*molecular aggregations of various kinds. The equilibrium between union and disunion of the amino acids that make up a protein is all to the advantage of disunion, but the aggregation of the protein with itself or other molecules might swing the equilibrium in the opposite direction: perhaps by removing the protein from access to the water which would be required to disintegrate it or by providing some particularly stable type of molecular association.

In such a scheme the protein appears only as a transient intermediate, an unstable way-station, which can either fall back to a mixture of its constituent amino acids or enter into the formation of a complex structural aggregate: amino acids $\rightleftharpoons$ protein $\rightarrow$ aggregate.

Such molecular aggregates, of various degrees of material and architectural complexity, are indispensable intermediates between molecules and organisms. We have no need to try to imagine the spontaneous formation of an organism by one grand collision of its component molecules. The whole process must

be gradual. The molecules form aggregates, small and large. The aggregates add further molecules, thus growing in size and complexity. Aggregates of various kinds interact with one another to form still larger and more complex structures. In this way we imagine the ascent, not by jumps or master strokes, but gradually, piecemeal, to the first living organisms.

First Organisms

Where may this have happened? It is easiest to suppose that life first arose in the sea. Here were the necessary salts and the water. The latter is not only the principal component of organisms, but prior to their formation provided a medium which could dissolve molecules of the widest variety and ceaselessly mix and circulate them. It is this constant mixture and collision of organic molecules of every sort that constituted in large part the "trials" of our earlier discussion of probabilities.

The sea in fact gradually turned into a dilute broth, sterile and oxygen-free. In this broth molecules came together in increasing number and variety, sometimes merely to collide and separate, sometimes to react with one another to produce new combinations, sometimes to aggregate into multimolecular formations of increasing size and complexity.

What brought order into such complexes? For order is as essential here as composition. To form an organism, molecules must enter into intricate designs and connections; they must eventually form a self-repairing, self-constructing dynamic machine. For a time this problem of molecular arrangement seemed to present an almost insuperable obstacle in the way of imagining a spontaneous origin of life, or indeed the laboratory

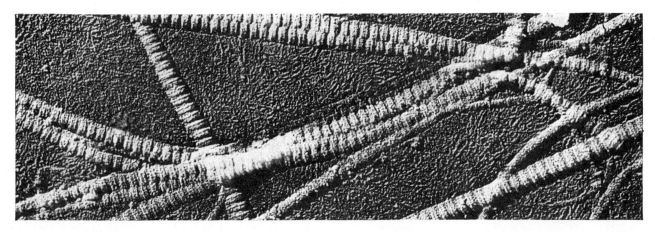

FIBRILS OF COLLAGEN formed spontaneously out of filaments such as those shown on the opposite page when 1 per cent of sodium chloride was added to the dilute acetic acid. These long fibrils are identical in appearance with those of collagen before dispersion.

synthesis of a living organism. It is still a large and mysterious problem, but it no longer seems insuperable. The change in view has come about because we now realize that it is not altogether necessary to *bring* order into this situation; a great deal of order is implicit in the molecules themselves.

The epitome of molecular order is a crystal. In a perfect crystal the molecules display complete regularity of position and orientation in all planes of space. At the other extreme are fluids—liquids or gases—in which the molecules are in ceaseless motion and in wholly random orientations and positions.

Lately it has become clear that very little of a living cell is truly fluid. Most of it consists of molecules which have taken up various degrees of orientation with regard to one another. That is, most of the cell represents various degrees of approach to crystallinity—often, however, with very important differences from the crystals most familiar to us. Much of the cell's crystallinity involves molecules which are still in solution—so-called liquid crystals—and much of the dynamic, plastic quality of cellular structure, the capacity for constant change of shape and interchange of material, derives from this condition. Our familiar crystals, furthermore, involve only one or a very few types of molecule, while in the cell a great variety of different molecules come together in some degree of regular spacing and orientation—*i.e.*, some degree of crystallinity. We are dealing in the cell with highly mixed crystals and near-crystals, solid and liquid. The laboratory study of this type of formation has scarcely begun. Its further exploration is of the highest importance for our problem.

In a fluid such as water the molecules are in very rapid motion. Any molecules dissolved in such a medium are under a constant barrage of collisions with water molecules. This keeps small and moderately sized molecules in a constant turmoil; they are knocked about at random, colliding again and again, never holding any position or orientation for more than an instant. The larger a molecule is relative to water, the less it is disturbed by such collisions. Many protein and nucleic acid molecules are so large that even in solution their motions are very sluggish, and since they carry large numbers of electric charges distributed about their surfaces, they tend even in solution to align with respect to one another. It is so that they tend to form liquid crystals.

We have spoken above of architectonic tendencies even among some of the relatively small molecules: the lecithins and cephalins. Such molecules are insoluble in water yet possess special groups which have a high affinity for water. As a result they tend to form surface layers, in which their water-seeking groups project into the water phase, while their water-repelling portions project into the air, or into an oil phase, or unite to form an oil phase. The result is that quite spontaneously such molecules, when exposed to water, take up highly oriented positions to form surface membranes, myelin figures and other quasi-crystalline structures.

Recently several particularly striking examples have been reported of the spontaneous production of familiar types of biological structure by protein molecules. Cartilage and muscle offer some of the most intricate and regular patterns of structure to be found in organisms. A fiber from either type of tissue presents under the electron microscope a beautiful pattern of cross striations of various widths and densities, very regularly spaced. The proteins that form these structures can be coaxed into free solution and stirred into completely random orientation. Yet on precipitating, under proper conditions, the molecules realign with regard to one another to regenerate with extraordinary fidelity the original patterns of the tissues [*see illustration above*].

We have therefore a genuine basis for the view that the molecules of our oceanic broth will not only come together spontaneously to form aggregates but in doing so will spontaneously achieve various types and degrees of order. This greatly simplifies our problem. What it means is that, given the right molecules, one does not have to do everything for them; they do a great deal for themselves.

Oparin has made the ingenious suggestion that natural selection, which Darwin proposed to be the driving force of organic evolution, begins to operate at this level. He suggests that as the molecules come together to form colloidal aggregates, the latter begin to compete with one another for material. Some aggregates, by virtue of especially favorable composition or internal arrangement, acquire new molecules more rapidly than others. They eventually emerge as the dominant types. Oparin suggests further that considerations of optimal size enter at this level. A growing colloidal particle may reach a point at which it becomes unstable and breaks down into smaller particles, each of which grows and redivides. All these phenomena lie within the bounds of known processes in nonliving systems.

The Sources of Energy

We suppose that all these forces and factors, and others perhaps yet to be revealed, together give us eventually the first living organism. That achieved, how does the organism continue to live?

We have already noted that a living organism is a dynamic structure. It is the site of a continuous influx and outflow of matter and energy. This is the very sign of life, its cessation the best evidence of death. What is the primal organism to use as food, and how derive the energy it needs to maintain itself and grow?

For the primal organism, generated under the conditions we have described, only one answer is possible. Having arisen in an oceanic broth of organic molecules, its only recourse is to live upon them. There is only one way of doing that in the absence of oxygen. It is called fermentation: the process by which organisms derive energy by breaking organic molecules and rearranging their parts. The most familiar example of such a process is the fermentation of sugar by yeast, which yields alcohol as one of the products. Animal cells also ferment sugar, not to alcohol but to lactic acid. These are two examples from a host of known fermentations.

The yeast fermentation has the following over-all equation: $C_6H_{12}O_6 \rightarrow 2 CO_2 + 2 C_2H_5OH +$ energy. The result of fragmenting 180 grams of sugar into 88 grams of carbon dioxide and 92 grams of alcohol is to make available about 20,000 calories of energy for the use of the cell. The energy is all that the cell derives by this transaction; the carbon dioxide and alcohol are waste products which must be got rid of somehow if the cell is to survive.

The cell, having arisen in a broth of organic compounds accumulated over the ages, must consume these molecules by fermentation in order to acquire the energy it needs to live, grow and reproduce. In doing so, it and its descendants are living on borrowed time. They are consuming their heritage, just as we in our time have nearly consumed our heritage of coal and oil. Eventually such a process must come to an end, and with that life also should have ended. It would have been necessary to start the entire development again.

Fortunately, however, the waste product carbon dioxide saved this situation. This gas entered the ocean and the atmosphere in ever-increasing quantity. Some time before the cell exhausted the supply of organic molecules, it succeeded in inventing the process of photosynthesis. This enabled it, with the energy of sunlight, to make its own organic molecules: first sugar from carbon dioxide and water, then, with ammonia and nitrates as sources of nitrogen, the entire array of organic compounds which it requires. The sugar synthesis equation is: $6 CO_2 + 6 H_2O +$ sunlight $\rightarrow C_6H_{12}O_6 + 6 O_2$. Here 264 grams of carbon dioxide plus 108 grams of water plus about 700,000 calories of sunlight yield 180 grams of sugar and 192 grams of oxygen.

This is an enormous step forward. Living organisms no longer needed to depend upon the accumulation of organic matter from past ages; they could make their own. With the energy of sunlight they could accomplish the fundamental organic syntheses that provide their substance, and by fermentation they could produce what energy they needed.

Fermentation, however, is an extraordinarily inefficient source of energy. It leaves most of the energy potential of organic compounds unexploited; consequently huge amounts of organic material must be fermented to provide a modicum of energy. It produces also various poisonous waste products—alcohol, lactic acid, acetic acid, formic acid and so on. In the sea such products are readily washed away, but if organisms were ever to penetrate to the air and land, these products must prove a serious embarrassment.

One of the by-products of photosynthesis, however, is oxygen. Once this was available, organisms could invent a new way to acquire energy, many times as efficient as fermentation. This is the

EXPERIMENT of S. L. Miller made amino acids by circulating methane (CH_4), ammonia (NH_3), water vapor (H_2O) and hydrogen (H_2) past an electrical discharge. The amino acids collected at the bottom of apparatus and were detected by paper chromatography.

BIBLICAL ACCOUNT of the origin of life is part of the Creation, here illustrated in a 16th-century Bible printed in Lyons. On the first day (*die primo*) God created heaven and the earth. On the second day (*die secundo*) He separated the firmament and the waters. On the third day (*die tertio*) He made the dry land and plants. On the fourth day (*die quarto*) He made the sun, the moon and the stars. On the fifth day (*die quinto*) He made the birds and the fishes. On the sixth day (*die sexto*) He made the land animals and man. In this account there is no theological conflict with spontaneous generation. According to *Genesis* God, rather than creating the animals and plants directly, bade the earth and waters bring them forth. One theological view is that they retain this capacity.

process of cold combustion called respiration: $C_6H_{12}O_6 + 6 O_2 \rightarrow 6 CO_2 + 6 H_2O$ + energy. The burning of 180 grams of sugar in cellular respiration yields about 700,000 calories, as compared with the approximately 20,000 calories produced by fermentation of the same quantity of sugar. This process of combustion extracts all the energy that can possibly be derived from the molecules which it consumes. With this process at its disposal, the cell can meet its energy requirements with a minimum expenditure of substance. It is a further advantage that the products of respiration—water and carbon dioxide—are innocuous and easily disposed of in any environment.

Life's Capital

It is difficult to overestimate the degree to which the invention of cellular respiration released the forces of living organisms. No organism that relies wholly upon fermentation has ever amounted to much. Even after the advent of photosynthesis, organisms could have led only a marginal existence. They could indeed produce their own organic materials, but only in quantities sufficient to survive. Fermentation is so profligate a way of life that photosynthesis could do little more than keep up with it. Respiration used the material of organisms with such enormously greater efficiency as for the first time to leave something over. Coupled with fermentation, photosynthesis made organisms self-sustaining; coupled with respiration, it provided a surplus. To use an economic analogy, photosynthesis brought organisms to the subsistence level; respiration provided them with capital. It is mainly this capital that they invested in the great enterprise of organic evolution.

The entry of oxygen into the atmosphere also liberated organisms in another sense. The sun's radiation contains ultraviolet components which no living cell can tolerate. We are sometimes told that if this radiation were to reach the earth's surface, life must cease. That is not quite true. Water absorbs ultraviolet radiation very effectively, and one must conclude that as long as these rays penetrated in quantity to the surface of the earth, life had to remain under water. With the appearance of oxygen, however, a layer of ozone formed high in the atmosphere and absorbed this radiation. Now organisms could for the first time emerge from the water and begin to populate the earth and air. Oxygen provided not only the means of obtaining adequate energy for evolution but the protective blanket of ozone which alone made possible terrestrial life.

This is really the end of our story. Yet not quite the end. Our entire concern in this argument has been to bring the origin of life within the compass of natural phenomena. It is of the essence of such phenomena to be repetitive, and hence, given time, to be inevitable.

This is by far our most significant conclusion—that life, as an orderly natural event on such a planet as ours, was inevitable. The same can be said of the whole of organic evolution. All of it lies within the order of nature, and apart from details all of it was inevitable.

Astronomers have reason to believe that a planet such as ours—of about the earth's size and temperature, and about as well-lighted—is a rare event in the universe. Indeed, filled as our story is with improbable phenomena, one of the least probable is to have had such a body as the earth to begin with. Yet though this probability is small, the universe is so large that it is conservatively estimated at least 100,000 planets like the earth exist in our galaxy alone. Some 100 million galaxies lie within the range of our most powerful telescopes, so that throughout observable space we can count apparently on the existence of at least 10 million million planets like our own.

What it means to bring the origin of life within the realm of natural phenomena is to imply that in all these places life probably exists—life as we know it. Indeed, I am convinced that there can be no way of composing and constructing living organisms which is fundamentally different from the one we know—though this is another argument, and must await another occasion. Wherever life is possible, given time, it should arise. It should then ramify into a wide array of forms, differing in detail from those we now observe (as did earlier organisms on the earth) yet including many which should look familiar to us—perhaps even men.

We are not alone in the universe, and do not bear alone the whole burden of life and what comes of it. Life is a cosmic event—so far as we know the most complex state of organization that matter has achieved in our cosmos. It has come many times, in many places—places closed off from us by impenetrable distances, probably never to be crossed even with a signal. As men we can attempt to understand it, and even somewhat to control and guide its local manifestations. On this planet that is our home, we have every reason to wish it well. Yet should we fail, all is not lost. Our kind will try again elsewhere.

The Amateur Scientist

6

Conducted by C. L. Stong
January 1970

*Experiments in generating the constituents of living
matter from inorganic substances*

No one has created life in the laboratory. The possibility is vanishingly small that a viable organism can be compounded. Yet that slender chance continues to intrigue some experimenters, and occasionally progress is reported. For example, a particularly dramatic experiment that amateurs can perform was described 17 years ago by Stanley L. Miller, then one of Harold Urey's students at the University of Chicago. Miller circulated a sterile mixture of water vapor, methane, ammonia and hydrogen through an electric spark. At the beginning of the experiment the mixture was crystal clear. Within 24 hours the condensed liquid turned noticeably pink, and after a week it became deep red, turbid and slightly viscid. Chemical analysis disclosed that the fluid contained a number of organic substances, including several amino acids, which are the building blocks of proteins.

Miller had not compounded these substances by controlled chemical synthesis. He merely established a physical environment that favored their spontaneous generation. It was the kind of environment that might have existed on the primitive earth.

Other experimenters have since reported similar achievements. Heat and various forms of electromagnetic radiation have been substituted for Miller's energizing spark to generate various constituents of living organisms, including purines, pyrimidines, protein-like polymers and nucleic acid polymers. In addition techniques have been devised for encouraging some of these materials to unite spontaneously into ordered structures. As such experimentation continues there is a minute but growing probability that a structure will eventually appear that feeds on the nutrients in its surroundings and reproduces itself.

As the biologist George Wald has pointed out: "However improbable the event in a single trial, it becomes increasingly probable as the trials are multiplied. Eventually the event becomes virtually inevitable" [see "The Origin of Life," by George Wald; article 5, beginning on page 47].

Fascinated by this promise, the experimenters keep trying. Even amateurs are beginning to get in on the fun. One is Carl Fromer of Staten Island, N.Y., who is a premedical student at Columbia University. Fromer explains how to do Miller's experiment and discusses a few others:

"Any experiment that is designed to generate from inorganic substances the constituents of living organisms must provide two basic conditions in the reaction vessel: the vessel must be devoid of life and lacking in free oxygen. The first of these conditions was mentioned more than a century ago in an obscure letter by Charles Darwin: 'It is often said that all of the conditions for the first production of a living organism are now present, which could ever have been present. But if (And, oh, what a big if!) we could conceive in some warm little pond, with all sorts of ammonia and phosphoric salts, light, heat, electricity, etc., present, that a protein compound was chemically formed ready to undergo still more complex changes, at the present day such matter would be instantly devoured or absorbed, which would not have been the case before living creatures were formed!'

"The second basic condition, the absence of free oxygen, was first emphasized by the Russian biologist A. I. Oparin in his classic volume of 1936, *The Origin of Life*. Essentially Oparin pointed out that free oxygen, a highly reactive gas, would promptly combine with

and destroy all organic compounds of interest. The reaction vessel must be sterile and airtight.

"Miller's apparatus consisted of two glass bulbs interconnected by a pair of glass tubes in an O configuration so that it was a closed system. Vapor from boiling water in one bulb rose through the outlet tube and entered the second bulb. The outlet tube of the second bulb contained a spark gap and was cooled just below the spark gap by a surrounding water jacket. Condensation that formed in this region drained through a U trap to the boiler, ready for another trip through the apparatus. Reaction products that formed in the vicinity of the spark gap accumulated in the water. Miller's construction involves some rather tricky glassblowing. The configuration did not seem critical, and so I made a simplified version that works just about as well.

"My apparatus consists of a closed tube of Pyrex glass 3.5 centimeters in diameter and 60 centimeters long fitted with a homemade heating mantle, a cooling jacket and platinum-wire electrodes for the spark gap [see illustration on page 58]. Access to the interior is provided by a side arm of eight-millimeter Pyrex tubing supplied with a stopcock. The tube operates vertically, with the spark gap at the top. Vapor rises by convection from boiling water at the bottom. Some vapor circulates through the spark gap, where the enclosed gases react, condenses on the chilled wall below and trickles back to the boiler. As in Miller's apparatus, the reaction products accumulate in the water.

"The tube is homemade. I used an ordinary gas-air blowtorch for softening the glass. The air supply was enriched with oxygen to develop the temperature required for working Pyrex. If the ex-

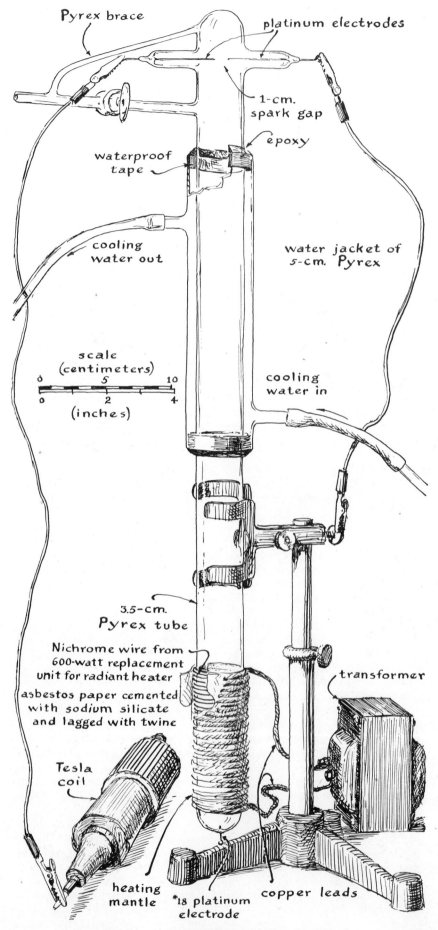

Labels on figure (clockwise from top):

Pyrex brace

platinum electrodes

1-cm. spark gap

epoxy

waterproof tape

cooling water out

water jacket of 5-cm. Pyrex

cooling water in

scale (centimeters)
0 5 10
0 2 4
(inches)

3.5-cm. Pyrex tube

Nichrome wire from 600-watt replacement unit for radiant heater

asbestos paper cemented with sodium silicate and lagged with twine

transformer

Tesla coil

heating mantle

#18 platinum electrode

copper leads

Carl Fromer's apparatus for generating amino acids

perimenter does not have access to glass-blowing facilities, arrangements could be made to have the tube assembled by a neon-sign shop.

"One end of the 35-millimeter tubing was collapsed in the flame and blown into a hemisphere. A small hole was then blown in the side, and the access tube was sealed to the hole. After welding a brace of glass rod to support the access tube, I collapsed the remaining open end of the 35-millimeter tubing and expanded it into a hemisphere by blowing into the access tube. The electrodes were sealed to a pair of opposing holes blown near the top of the tube and in the center of the hemispherical bottom. Platinum does not make a vacuum-tight seal with Pyrex, and so I dabbed epoxy cement around each wire at the point where it emerged from the glass.

"The water jacket consists of a 22-centimeter length of Pyrex tubing five centimeters in diameter equipped with eight-millimeter inlet and outlet tubes. The ends of the jacket were closed with rings of plastic tape wrapped around the 35-millimeter tube. The rings make a snug fit with the jacket. I sealed the rings to the jacket with a thick coating of epoxy cement.

"The heating mantle was made by winding a coil of Nichrome wire, along with asbestos insulation, around the lower end of the tube. A strip of asbestos paper 11 centimeters wide was cemented around the glass with sodium silicate (water glass). A flexible copper lead, insulated with asbestos, was crimped to one end of the Nichrome wire and lashed to the paper with twine near one edge. A layer of wire was wound over the paper at a turn spacing of five millimeters. Four layers of wire and paper were applied. After a flexible lead was attached to the outer end of the wire and lashed to the end of the final layer the coil was completed with a cover of asbestos paper lagged in place with twine and cemented.

"The finished tube was exhausted to the limit of a mechanical air pump. I used the compressor from a discarded electric refrigerator for an air pump, connecting the tube to the inlet of the compressor. The tube was then sterilized by electron bombardment. One terminal of a 15,000-volt, 60-milliampere transformer from a neon sign was connected to both spark-gap electrodes; the other terminal was attached to the electrode in the bottom of the tube. Increasing potential was gradually applied to the input terminals of the high-voltage transformer by a variable transformer until the tube filled with glow discharge. The

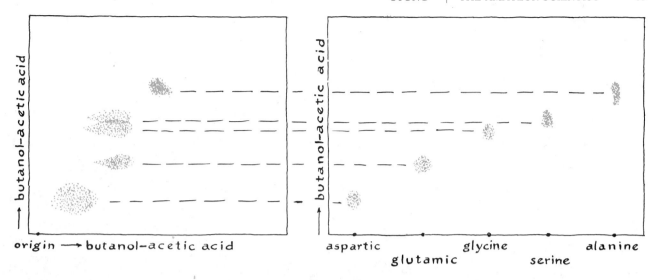

Chromatogram of known (right) *and unknown* (left) *amino acids*

tips of the electrodes became red-hot. The bombardment was continued for five hours, after which the stopcock was closed.

"The apparatus was transferred to a sterile chamber for filling, along with a pressure cooker containing about a liter of distilled water that had been boiled at a pressure of 15 pounds per square inch for 30 minutes. Some 500 milliliters of the cooled water was transferred to a sterile beaker. The access tube was sterilized briefly in a flame and, when it had cooled, was dipped into the water. Water rushed into the tube when the stopcock was opened. I filled the tube to the upper edge of the heating mantle, using about 100 milliliters of water.

"The experiments were made with gases of reagent grade that come compressed in lecture bottles. The gases and the other required materials, including the glass, are available from distributors of scientific supplies. The gases are mixed in the proportion of one part hydrogen and two parts each of anhydrous ammonia and methane (by volume). The proportions need not be exact. I had no precise gauges, and so I measured the gases with three toy rubber balloons that expand into approximate spheres.

"After the balloons had been sterilized in the pressure cooker one was coupled to the outlet of the hydrogen cylinder and filled to a diameter of six inches. The remaining two balloons were similarly filled to a diameter of 7.5 inches with ammonia and methane respectively. The gases were combined in one of the balloons by coupling, with a short glass tube, the balloon containing hydrogen to the one containing ammonia. The balloon containing hydrogen

was squeezed to force the gas into the ammonia. The methane was combined with the ammonia in the same way.

"The balloon now containing all three gases was coupled to the access tube of the apparatus and the stopcock was opened to admit the mixture. After the stopcock was closed and the balloon (still containing a substantial amount of gas) had been removed, an empty balloon was coupled to the access tube. When the stopcock was opened, the trapped gas expanded to atmospheric pressure. A more elegant technique for metering the gas can doubtless be devised, but this one works and is inexpensive.

"The filled tube was supported upright by an apparatus stand. After power was applied to the heater, the water came to a gentle boil in about two hours. In the meantime a Tesla coil was connected to one of the spark-gap electrodes and the companion electrode of the spark gap was grounded. When the water reached the boiling point, the coil was turned on and adjusted to an output potential just sufficient to jump the spark gap. Coils of this type are designed for locating leaks in vacuum systems made of glass and are available commercially.

"Within 24 hours the water assumed the color of weak tea, and in seven days it became dark brown. A brown waxlike deposit formed in the upper portion of the tube. After the power was turned off and the tube had cooled, I transferred the fluid under vacuum to a sterile flask. The flask had been fitted with a sterile rubber stopper containing two holes, one of which made a snug fit with the access tube of the apparatus. A hose attached the air pump to the other hole.

"The apparatus was placed in the horizontal position so that the fluid would drain into the flask when the air was pumped out. The fluid was concentrated by evaporation under vacuum in the flask to a volume of approximately 20 milliliters and transferred to a sterile test tube of known weight in which it was evaporated to dryness under vacuum. The net weight of the solid reaction products was 816 milligrams.

"The material was analyzed for amino acids by thin-layer chromatography and by an ion-exchange column. To make the chromatographic analysis I dissolved 25 milligrams of the dried material in 100 milliliters of distilled water to which 364 milligrams of hydrochloric acid was added. Similar solutions, each containing one of 20 known amino acids, were prepared by the same procedure and stored in separate, labeled containers.

"The specimens were applied, as small spots with a micropipette, to an Eastman Chromagram sheet coated with cellulose. To minimize the area covered by the spots and thus improve the resolution of the chromatogram I applied a third of a microliter of each specimen to the cellulose, dried the spot thoroughly and then made similar applications on top of the first one until one microgram had been deposited.

"The unknown material was placed near one corner of a Chromagram sheet at a distance of three centimeters from the edges. The known specimens were spotted in a row on another sheet, two centimeters apart and three centimeters from one edge of the sheet. Both sheets were then placed for development in a rectangular, closed glass dish that contained wash solution.

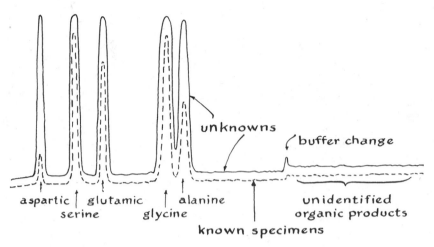

aspartic | glutamic | alanine | unidentified
serine | glycine | organic products
known specimens
unknowns | buffer change

Ion-exchange graph of five amino acids

"The wash solution was prepared by shaking together 100 milliliters each of distilled water, acetic acid and butanol. After standing for a few minutes this mixture separates. Fifty milliliters of the top layer was transferred to the rectangular container. The container was lined with a sheet of filter paper saturated with wash solution. The Chromagram sheets were inserted in the rectangular container with the specimens at the bottom.

"Wash solution migrates upward by capillary attraction and carries the acids along at characteristic rates that vary with the affinity of the acids for cellulose. Development was continued until the wash solution migrated to within five millimeters of the top edge. Both developed sheets were dried.

"The sheet containing the unknown substances was then redeveloped by turning it 90 degrees and again inserting it into the wash solution with the specimen at the bottom. The specimen substances migrated at right angles with respect to their first transit and separated still more. The dried chromatograms appeared blank to the eye, but the amino acids became visible as purple spots after the sheets were sprayed with ninhydrin solution and baked at 100 degrees Celsius in a kitchen oven for five minutes. I made up the solution by dissolving 200 milligrams of ninhydrin in 100 milliliters of acetone to which one drop of collidine was added immediately before use.

"Five spots of color appeared on the sheet of the unknown substances. The distance through which each spot migrated matched the migration distance of one acid of the sheet that had been spotted with known acids. A third sheet was spotted with these five known substances, developed and compared with the unknown substances. The unknown substances were identified as aspartic acid, glutamic acid, serine, glycine and alanine, in the approximate proportions of one aspartic acid, two glutamic acid, 15 glycine, 12 serine and 13 alanine [*see illustration on previous page*].

"At about the time the chromatograms were made a biochemist volunteered to run a sample of the material on an automatic ion-exchange analyzer. In this instrument compounds are washed through a vertical tube of glass packed with granular ion-exchange resins. The compounds migrate at characteristic rates and emerge from the column separately into a stream of clear fluid that reacts with the fractions to become colored.

"For the analysis of amino acids the clear fluid is ninhydrin solution. Color changes are monitored by a photoelectric cell that actuates a pen recorder. The emergence of each amino acid appears as a peak in the graph that is plotted automatically by the recorder. The machine includes several columns that can be operated simultaneously for plotting slightly displaced graphs of two or more mixtures, for example a known and an unknown mixture of amino acids. A graph of the amino acid mixture that had been synthesized by my apparatus, combined with a superposed graph of known amino acids, confirmed the chromatographic analysis and also indicated the proportionate yields mentioned above [*see illustration above*].

"The synthesized amino acids can be used to perform a fascinating experiment. About 10 years ago Sidney W. Fox of the Institute for Space Biosciences at Florida State University developed a simple technique that encourages the amino acids to unite spontaneously in the form of well-organized structures that he refers to as microspherules. Fox shared the opinion of some experimenters that amino acid molecules, which formed in the vaporous atmosphere of the primitive earth, tended to be carried into the oceans by rain. These compounds did not break down readily, and the oceans became an increasingly rich mixture of organic compounds. Fox reasoned that some of the compounds would occasionally be washed onto hot lava, where they would dry, cook and subsequently be washed again into the sea. How would the compounds respond to this sequence of events? Fox put the question to nature by an experiment. The answer turned out to be that the amino acids would organize themselves into microspherules!

"His experiment sounded interesting, and so I decided to try it. Hot lava is scarce on Staten Island, but I own a flat plate of ceramic that contains a number of hemispherical depressions about a centimeter in diameter and half as deep. I placed about 50 milligrams of the dried reaction product from my apparatus in each of four depressions and the same amount of a mixture (in one-to-one proportions) of commercially prepared aspartic and glutamic acids in four other depressions to serve as a control. The ceramic plate was put in an electric oven at a temperature of 85 degrees C. At the end of about an hour the powdered materials became a highly viscous mass resembling dark brown tar.

"I then filled each depression with a 1 percent solution of salt water to simulate the concentration of sodium chloride in the primordial sea and thereafter, for six hours, replaced the evaporating fluid from time to time with distilled water. Finally, with a pipette I transferred the end products (the synthesized amino acids and the control) to separate centrifuge tubes and rotated them for 25 minutes at 2,600 revolutions per minute. Smears of the dense fraction of both specimens that settled to the bottom of the tubes were examined with a microscope. Both contained microspherules that were identical in appearance, but the population in the control specimen was substantially larger. The accompanying photomicrograph [*top of opposite page*] depicts microspherules made with the synthesized amino acids.

"Microspherules appear to consist of an inner core enclosed by a transparent membrane. A careful study of the structures would require the use of a phase-contrast microscope. I do not have one, and so I stained a smear of the specimen with bromophenol blue and examined it with a conventional microscope at a magnification of 1,000 diameters. For some minutes following their preparation microspherules appear to be jellylike in consistency and somewhat tacky. Later they solidify and retain their spherical shape for weeks.

"Fox demonstrated that freshly made microspherules tend to unite into long chainlike bodies that resemble a string of microscopic beads. To encourage the formation of these structures Fox applied slight pressure to the freshly cooked mixture. I made them by first depositing a thin ring of a cement known as gold size on a microscope slide. In the resulting shallow cavity I placed a drop of freshly made microspherule solution. The cavity was closed with a cover slip. When pressure was applied to the cover slip by the point of a needle, the microspherules linked into chains [see bottom illustration at right].

"Fox has also demonstrated that various organic compounds, including amino acids, can be generated by exposing a mixture of the same gases to heat. His apparatus resembles Miller's in that it consists of a closed loop containing a cooling jacket and a pair of bulbs, one of which serves as a boiler. For the spark gap Fox substituted a reaction vessel of quartz that contains hot sand. Unreacted gases flow through a Pyrex tube to the bottom of the vessel and migrate upward through the sand to an outlet tube for another transit through the apparatus. The sand is maintained at a temperature of 900 to 1,100 degrees C. by an electric furnace that surrounds the quartz reaction vessel [see illustration on page 62].

"The gases must be forced through the sand by pressure, which is developed by an aspirator that Fox inserted in the system immediately above the boiler. Condensed fluid is admitted to the boiler periodically by opening a stopcock in the return line. So far Fox has produced with quartz sand 13 amino acids, including aspartic acid, threonine, serine, glutamic acid, proline, glycine, alanine, valine, alloisoleucine, isoleucine, leucine, tyrosine and phenylalanine. Twelve of the substances have also been produced by packing the reaction vessel with silica gel. On the other hand, b-alanine, sarcosine and N-methylalanine have not been synthesized by heat although they appear in Miller's apparatus.

"In a typical experiment Fox passes methane gas through concentrated aqueous ammonia and transfers the resulting mixture to the apparatus. After the gases have circulated through the heated sand

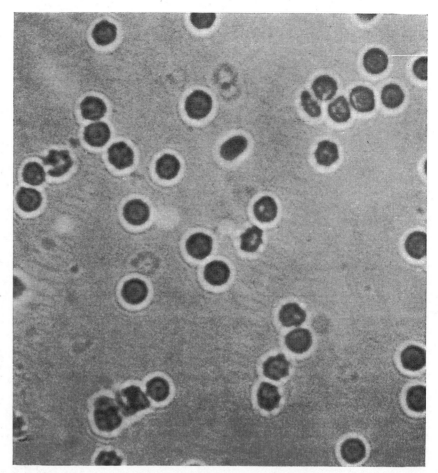

Microspherules of synthesized amino acids

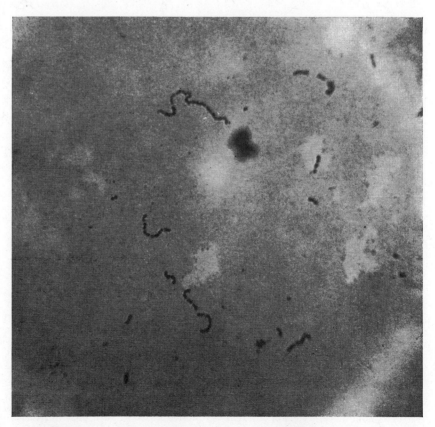

Chainlike structures formed by microspherules of synthesized amino acids

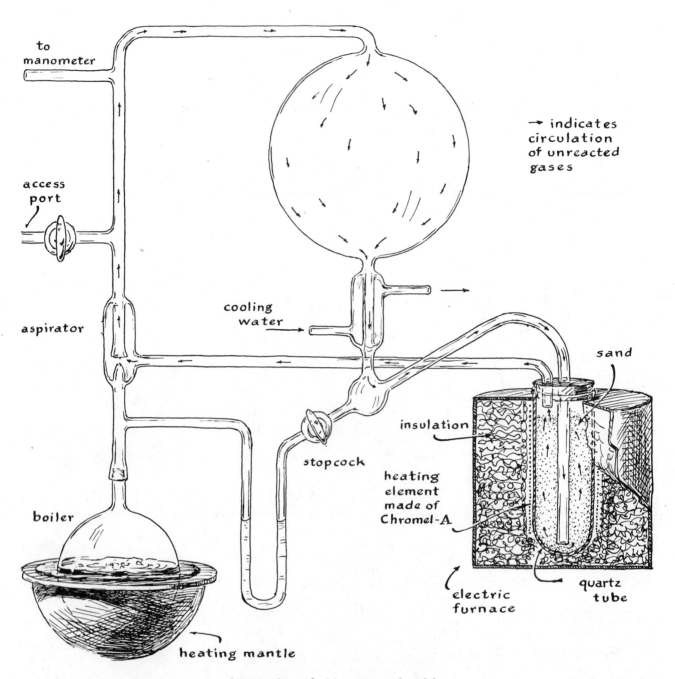

Apparatus for synthesizing amino acids with heat

for various periods of time that may range from minutes to hours, they are transferred to a closed flask and absorbed by a cold solution of 3 N aqueous ammonia. The cold solution is then heated to 75 degrees C. for various lengths of time in a closed flask and finally evaporated under vacuum. A solution of 4 N hydrochloric acid is circulated through the dried residue (refluxed) for five hours.

"The resulting hydrolysate is evaporated to dryness under vacuum and dissolved in a citrate buffer at pH 2.2. The mixture is then ready for analysis and use. Fox found that the yield of different amino acids varies with the materials in the reaction vessel and with the tem-

perature of the vessel. For example, when the reaction vessel is packed with quartz sand maintained at a temperature of 950 degrees C., aspartic acid constitutes 3.4 percent of the final product, but when silica gel is used at the same temperature, the production falls to 2.5 percent. When the temperature of the silica gel is raised to 1,050 degrees, the production of aspartic acid rises to 15.2 percent but the production of glycine falls from 68.8 percent to 24.4 percent.

"In undertaking these experiments I made a point of keeping the hazards in mind. Hydrogen and methane form explosive mixtures with air. Anhydrous ammonia is toxic and can burn the skin. Ideally the gases should be handled in a

fume hood. I filled the balloons and transferred the gases to the apparatus in my backyard. Ammonia migrates slowly through rubber. If you squeeze the balloon with your bare hands, the gas may burn your palms. I wore neoprene gloves. The cooling jacket prevents explosive steam pressure from building up inside the apparatus, *but only if a continuous stream of cold water circulates through the jacket.*

"As time permits I intend to continue the experimentation by including phosphoric acids in the mixture, in the hope of synthesizing some of the nucleic acid polymers and perhaps inducing them to combine with the protein-like bodies."

III

PROTOCELLS AND FOSSILS

III PROTOCELLS AND FOSSILS

INTRODUCTION

The fossils most of us are accustomed to are visible with the unaided eye in sedimentary rocks. These are the remains of the hard shells and skeletons of animals and of the more enduring parts of plants. Less frequently, casts of soft-bodied animals, leaves, and even worm tubes are preserved in the fossil record. All these represent relics of a biology of multicellular creatures. The oldest known fossils of this type are those of sponges and trilobites, 600 m yr in age. No multicellular fossils have been found in sedimentary rocks older than these, although similar rock formations extend backward in time for another 3200 m yr. But other, more subtle kinds of fossils are seen in older rocks.

The vast expanse of geological time is difficult to appreciate. From 4500 m yr ago, when the Earth's solid crust formed, to the first appearance of sedimentary rocks 3800 m yr ago, we have a 700 m yr period of time during which volcanoes discharged gases and early cellular life came into being and evolved. This period is as great as the entire span of the Earth's multicellular fossil record—a period ranging from the first worm to the present-day complex ecology. This earliest time has left few direct hints of its nature, because all rocks this old have been reworked, weathered, crushed, and heated innumerable times.

A second, even greater block of time stretches from 3800 m yr to 600 m yr ago—a total of 3200 m yr to which we do have access, even though indirect. The key to study of this period is analysis of the ancient sedimentary rock column. Both articles in this section report how this analysis might be accomplished. An enormous amount of laboratory work remains to be done.

Geoffrey Eglinton and Melvin Calvin show how organic compounds can be extracted from the matrix of rocks, separated, and identified by modern analytical chemical techniques. If the rocks are chosen carefully to avoid cracked, weathered, or contaminated specimens, the assumption can be made that organic compounds entrained within the matrix are indigenous. Certainly, many organic molecules will have been altered or degraded during the period of their capture. However, certain classes of compounds are very stable, such as the isoprenoid residues of chlorophyll; others degrade to characteristic stable products, such as the stable alkane fraction which is derived from lipids.

A biology uses characteristic sets of organic compounds. Porphyrins form part of the chlorophyll molecule of photosynthesizing plants and are used in all cells for energetic metabolism. These very stable, easily detected molecules could be considered one of many useful "flags" for a biological origin of the carbon compounds found in ancient rocks. As another example, long chain lipids of all organisms degrade to yield hydrocarbons characteristically enriched in isoprenoids and even-numbered straight-chain hydrocarbons. Hence the term "chemical fossils" is an apt one: the presence of these molecules in rocks is evidence for a flourishing biology.

The carbon isotope ratios of organic matter trapped within rocks can also be measured. Biological photosynthetic systems act essentially as fast one-way catchers of carbon dioxide, which is converted into organic carbon. Incorporation of the lighter isotope carbon-12 is favored slightly over the heavier carbon-13. Thus organic material formed by primitive cells during ancient times would be expected to show a distinct enrichment in carbon-12. It does.

Elso Barghoorn collected rocks from Precambrian sedimentary deposits such as the Swaziland Sequence of South Africa. This massive, well-preserved formation is about 3300 m yr old. Similar sources of ancient sediments abound in Canada, Australia, Greenland, South America, and India. The rocks were cut into ultra-thin sections, polished, and examined under the microscope. A surprisingly wide variety of microstructures could be seen. Bacterialike objects were found in the 3300 m yr old Fig Tree chert. In addition, spherical structures resembling modern coccal forms of photosynthetic blue-green bacteria were present. Other rocks of the North American Precambrian Shield, 2200 m yr old, yielded more complex spherical and filamentous microfossils. A whole new field of micropaleontology was opened.

From these ancient fossils we can infer that cell-like structures were present some 3200 m yr ago. A definite evolution toward complexity of form is visible in less ancient (only 2200 m yr old!) rocks. Can we conclude that a cellular biology was present 3200 m yr ago? Not really. A set of adequate criteria is sorely needed that would clearly distinguish the biological from the nonbiological.

For example, Barghoorn's oldest structures (3200 m yr) resemble not only cells but also organic microstructures which have been synthesized in the author's laboratory in modified Miller-Urey spark discharge experiments. The oldest unambiguous microfossil assemblage now known is the remains of filamentous blue-green bacteria reported by Dr. Lois Nagy of the University of Arizona.

Could carbon-isotope ratio disproportion serve as a biological flag? At first sight, yes. But *any* process that fixes carbon rapidly into organic material will also favor carbon-12 enrichment. Much more quantitative work remains to to done.

Could homogeneity of microfossil size distribution convince us that a fossil was biological in origin? Probably not, since we would expect more primitive cells to be less controlled in this and other aspects of their biosynthesis.

The current picture is that the earliest structures, 3200 m yr old, could be cells or protocells. After 1000 m yr of further evolution we find unambiguous microfossils 2200 m yr old. A great deal of work remains to elucidate this 1000 m yr period of cellular evolution, but the path is clear.

Suggested Further Reading

Brooks, J., and G. Shaw. 1973. *Origin and Development of Living Systems.* Academic Press, New York.

Ponnamperuma, Cyril (ed.). 1977. *Chemical Evolution of the Early Precambrian.* Academic Press, New York.

7

The Oldest Fossils

by Elso S. Barghoorn
May 1971

The remains of ancient bacteria and algae, some of them more than three billion years old, have been found in Africa, Australia and Canada. They provide evidence on the earliest stages of evolution

How did life originate on the earth? It was not so long ago that attempts to answer this question defined more areas of uncertainty than of agreement. Today a fossil record that was virtually unknown before the 1950's has been found to bear witness to three of the key events in the earliest stages of organic evolution. The fossils come from widely separated parts of the earth. All are preserved in unusual rocks of the Precambrian, the first and by far the longest interval in geologic history. The earliest of the fossils are more than three billion years old.

All that is known or conjectured about the terrestrial origin of life suggests that the first appearance of living organisms was preceded by the gradual development of a complex chemical environment. This environment is usually pictured as a kind of primordial broth, filled with such "organic" molecules as amino acids, sugars and other biologically important substances that came into existence through nonorganic processes. Millions of years must have been required for the accumulation, elaboration and differentiation of the broth. That period may be called the time of chemical evolution, a concept that owes much to the work of a leading student of abiotic synthesis, Cyril A. Ponnamperuma of the Ames Research Center of the National Aeronautics and Space Administration. As a preliminary stage in the total process of organic evolution, chemical evolution of course reaches its climax when lifeless organic molecules are assembled by chance into a living organism. This first form of life is what the Russian biochemist A. I. Oparin calls a "protobiont."

Judging from the various forms of life we know today, the first protobionts were probably microscopic in size and single-celled in structure, perhaps resembling the modern coccoid, or spherical, bacteria. Rather than imagining some specific organism, however, let us consider this first form of life more abstractly and give it a name that describes how it lives rather than how it looks. Call it a heterotroph, which is to say an organism that cannot manufacture all its own nutrients but must feed on organic molecules in the broth that surrounds it. (This assumes that the organism is immersed in an aqueous medium or at least rests on a wet surface; a supply of water is essential to protoplasmic life.) It seems reasonable to suppose that the organism was a heterotroph; to have at this stage a full-fledged autotroph, an organism that can manufacture its organic nutrients out of inorganic substances, would be asking too much of chance.

We can, however, demand that heterotrophs rather promptly give rise to autotrophs. Otherwise, as Preston Cloud of the University of California at Santa Barbara puts it, once the heterotrophs had "gobbled up all the goodies" within reach they would die off, making another accident of biosynthesis necessary in order to get things going again.

One way or another, then, autotrophs must evolve from an original heterotrophic population. This event in the development of life marks the time when the organic nutrients of the primordial broth are depleted and photosynthesis—the most plausible form of organic self-nutrition—must have been invented. Such an assumption is not based solely on logical considerations of thermodynamics and physiology. It is also supported by geological evidence that early organisms depended on photosynthesis to sustain themselves. Among the most ancient formations in the geological record are some showing signs that small amounts of free oxygen—the gaseous byproduct of photosynthesis—were present early in the history of the earth. This chemical evidence coincides with indications in the fossil record that some of the most primitive forms of terrestrial life, organisms that resembled modern bacteria and blue-green algae, were then increasing in numbers and diversity.

Many modern bacteria are photosynthetic, although they do not produce free oxygen; all modern blue-green algae are photosynthetic and all produce free oxygen. The evidence suggesting that the first autotrophs were organisms of the same primitive kind is significant for an additional reason. Bacteria and blue-green algae are alone among living things in the simplicity of their cells. They have neither a membrane-enclosed nucleus nor such specialized cellular organelles as mitochondria. Their genetic material is diffused throughout the cell, and they are incapable of either mitosis (body-cell division) or meiosis (germ-cell division). Both kinds of cell division require that the genetic material be organized into chromosomes.

Bacteria and blue-green algae are thus fundamentally different from other organisms: all other plants, all animals and the many forms that are neither entirely plant nor entirely animal. These other organisms have cells with nuclei and specialized organelles or specialized intracellular structures; they are called eukaryotic (truly nucleated), whereas bacteria and blue-green algae are prokaryotic (prenuclear). It would be surprising if the autotrophs on the lowest rungs of the evolutionary ladder were anything but prokaryotic.

The prokaryotic cell is notable for still another reason. Any organism whose genetic material is diffused throughout the

OLDEST KNOWN BACTERIUM, one of two primitive forms of life preserved in a Precambrian rock formation in South Africa, appears as a raised rectangular shape in this electron micrograph. What is seen is a carbon replica of the polished surface of a rock sample, shadowed with heavy metal. The fossil bacteria come from cherts of the Fig Tree formation; they are from .5 to .75 micron long and about .25 micron wide. The organisms are some 3.2 billion years old and have been given the name *Eobacterium isolatum*.

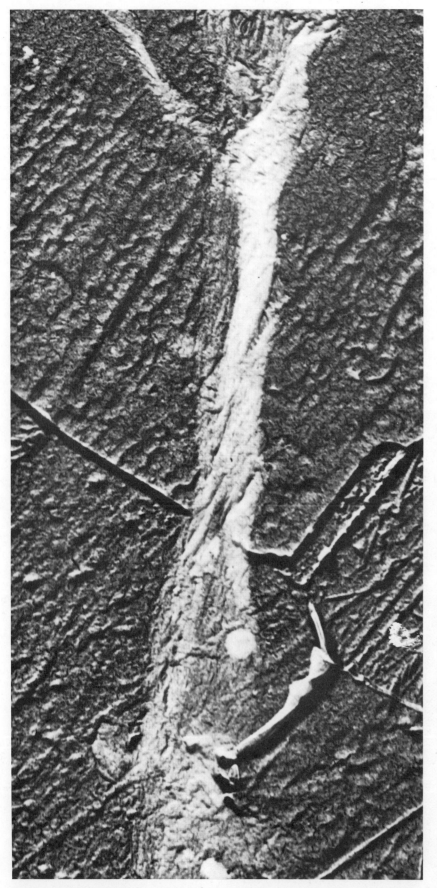

THREADLIKE FILAMENT of organic matter resembling decomposed plant tissue is another kind of fossil that appears in electron micrographs of the Fig Tree cherts. Some specimens are nine microns long. Not identifiable with any known organism, the filaments might conceivably be polymerized abiotic molecules from the "primordial broth."

cell and whose reproduction does not involve a recombination of parental genes is genetically conservative. In such an organism random mutations, instead of being preserved when they benefit the organism, tend to be damped out in a few generations. The blue-green algae are an outstanding example of such genetic conservatism. Many living species are almost indistinguishable in structure from species that flourished a billion or more years ago.

Against this background of fact and conjecture, how many of the events in the early stages of organic evolution can be labeled as being outstanding? There appear to be three such events, each of them a kind of threshold-crossing. The first, obviously the *sine qua non*, is successful biosynthesis: a crossing of the threshold separating the initial period of abiotic chemical evolution from the subsequent organic period. Perhaps someday fossil evidence of the earth's earliest organisms, the heterotrophs that crossed this first threshold, will be discovered. In the meantime the remains of their successors, the early photosynthetic autotrophs, provide the necessary proof that the first threshold had been passed.

The next threshold can be characterized as the threshold of diversification. As evolution progresses the first photosynthetic organism should not be limited to a few similar forms. Instead they should develop differences in shape and structure indicative of roles in a variety of ecological niches.

The third threshold divides prokaryotic organisms from all others. A world populated solely by bacteria and blue-green algae is conceivable; indeed, that is apparently the way it was. Viewed from our present vantage such a world seems poor in possibilities for further evolution. The extraordinary variety of plant and animal life that has arisen on the earth over the past 600 million years is due entirely to the invention of the eukaryotic cell, with its potential for genetic diversity.

Thanks to a lucky accident of fossil preservation, we now have evidence that each of these thresholds was successfully crossed during the vast span of Precambrian times. The Precambrian represents nearly four billion of the earth's first 4.5 billion years. It began with the formation of the earth and ended some 600 million years ago with the dawn of the Paleozoic era [*see illustration on page 77*]. Nothing is known of the first billion years or so; the world's oldest-known rocks, found in Africa, are not much more than three billion years old.

Precambrian rocks are found not only in Africa but also on every other continent. The best-known areas are the Canadian Shield in North America and the Fenno-Scandian Shield in Europe, but more than a third of Australia is also a Precambrian shield and there are sizable Precambrian areas in South America and Asia. Some Precambrian rocks are of igneous origin and some are of sedimentary origin. Most of the sediments have been heavily metamorphosed: changed in form and chemical composition by heat and pressure. In these metamorphic rocks not only fossils but also the faintest traces of organic matter have been obliterated.

A few Precambrian sediments have escaped substantial alteration. Extensive deposits of black shale, black chert and other stratified sediments are scattered through the major shield areas in virtually unmetamorphosed condition. These carbon-rich rocks—for example the formations in the Lake Superior region of North America, in the Transvaal of South Africa and in parts of western Australia—look even to the experienced eye very much like certain sediments of the Carboniferous epoch that are a mere 300 million years old.

The oldest-known group of Precambrian sediments is located in the border region between the Republic of South Africa and Swaziland. The formation is called the Swaziland Sequence; its stratified rocks are thousands of feet thick. One series of strata in the middle of the sequence, known as the Fig Tree formation, is well exposed in the Barberton Mountain Land, a gold-mining district near the town of Barberton in the eastern Transvaal. The Fig Tree rocks consist of black, gray and greenish cherts, interbedded with jaspers, ironstones, slates, shales and graywackes. In places the chert beds are 400 feet thick. The chert is usually fractured, but it is cemented together and the fractures are filled with quartz; parts of the formation show little evidence of metamorphism. The Fig Tree cherts contain traces of organic matter and a few microfossils.

The Barberton Mountain Land has long been an active mining area, and its geology has been studied in considerable detail. The age of the graywackes and shales has been determined independently by several laboratories using a method based on the decay of radioactive strontium and rubidium. This particular radioactive clock started to run 3.1 billion years ago, but there is good reason to believe the sediments were deposited

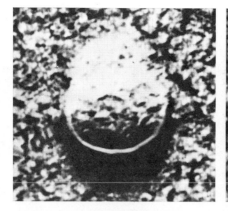 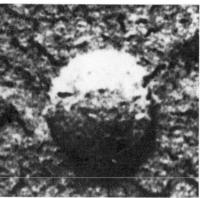

TWO CROSS SECTIONS of *Eobacterium* are seen in electron micrographs of metal-shadowed carbon replicas. In the fossil at left the outer and inner layers of the cell wall are visible. The wall, .015 micron thick, resembles the wall of living bacteria of the bacillus type.

somewhat earlier. Recently rocks lying close to the base of the Swaziland Sequence (and thus well below the level of the Fig Tree cherts) were shown to be 3.36 billion years old. In light of these age determinations it seems probable that the age of the Fig Tree cherts is in excess of 3.2 billion years.

In 1965 I collected cherts from several localities in the Fig Tree series, and samples were prepared for examination in my laboratory at Harvard University that summer. Two techniques were employed: thin sections of chert were cut for examination by reflected and transmitted white light under the microscope, and carbon replicas of etched and polished chert surfaces were made. The carbon replicas were then "shadowed" with metal and examined by transmission electron microscopy. J. William Schopf joined me in the study of the specimens.

When Schopf and I examined the thin sections under the light microscope, we could see that the rock matrix contained abundant laminations of dark-colored and virtually opaque organic matter. The

laminations were irregular but were usually aligned parallel to the strata of the chert, suggesting that they had originally been formed as part of an aqueous sedimentary deposit. No distortion was evident where the organic matter crossed the boundaries of the individual grains of chalcedony comprising the chert, which suggested to us that the process of deposition had emplaced the organic material within a silica-rich matrix before the silica was crystallized into chert. There was no evidence whatever that the silica was of secondary origin.

It was difficult to discern distinct bodies within the layers of organic matter under the light microscope. Our first success in isolating a Fig Tree organism was achieved with the carbon-replica technique; the electron microscope revealed a number of rod-shaped structures, preserved both in profile and in cross section. The rods are very small. They range in length from slightly under .5 micron to a little less than .7 micron, and in diameter from about .2 micron to a little more than .3 micron. In cross section the cell wall is sometimes seen to

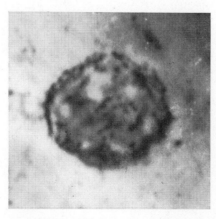 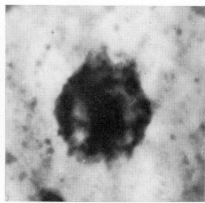

ALGA-LIKE SPHERES, seen in photomicrographs of thin sections of rock, are the other fossil organisms that are found in the Fig Tree formation. The diameter of the spheres is usually less than 20 microns. The organism is named *Archaeosphaeroides barbertonensis*.

consist of an inner and an outer layer, with a total wall thickness of .015 micron [*see top illustration on preceding page*]. This is comparable to the cell wall of many modern bacteria in both structure and dimension.

Electron microscopy also revealed the presence of organic material in the form of irregular, threadlike filaments lacking discernible structural detail. The threads are clearly native to the chert and not the products of contamination. They are as much as nine microns long and resemble decomposed plant material. Although these filaments are almost certainly of biological origin, they cannot be identified with any known type of organism. It has been suggested, probably as a result of wishful thinking, that they may be polymerized strands of abiotic organic matter from the primordial broth.

Schopf and I later succeeded in resolving larger microfossils in thin sections of Fig Tree chert under the light microscope. These fossils are spheroidal; measurements of 28 particularly well-defined specimens show that the majority are between 17 and 20 microns in diameter [*see bottom illustration on preceding page*]. Some have a darkened interior, as if cytoplasm within the spheroid had coalesced and become "coalified." Just as the rod-shaped organisms revealed by the electron microscope resemble certain modern bacteria, so the spheroids are not unlike some modern blue-green algae of the coccoid group. They may even be among the evolutionary precursors of such algae.

We have named the rods *Eobacterium isolatum*, a new genus and species. The generic name (*eo-* is the Greek root for "dawn") points to the great antiquity of the organism and to its bacterium-like appearance; the specific name defines its noncolonial, single-cell habit of growth. The spheroids we have named *Archaeosphaeroides barbertonensis*, also a new genus and species. Again the generic name refers to the organism's great age and its appearance; the specific name identifies its place of discovery. The existence of these two organisms, successful inhabitants of an aquatic environment more than three billion years ago, is evidence that the first evolutionary threshold—the transition from chemical evolution to organic evolution—had been safely crossed at some even earlier date. We now know that at least two living organisms appeared well before the first third of earth history had passed. If we accept the evidence (to which we shall return) that the alga-like Fig Tree organisms were photosynthetic, an important

geochemical event must also have occurred. With the onset of photosynthesis free oxygen would have begun to appear among the other constituents of the environment. The appearance of free oxygen was an event destined to have profound influence, both biological and geological, on the subsequent history of the earth.

Evidence of the second evolutionary threshold-crossing comes from North America. A remarkable outcropping of Precambrian rocks along the shore of Lake Superior in western Ontario shows a sequence of sediments known as the Gunflint Iron formation. The rocks at the base of the Gunflint formation include beds of black chert three to nine inches thick. The beds are exposed—in some places more or less continuously and in others as isolated outcrops—over a distance of some 115 miles in Ontario, from the vicinity of Schreiber on the east to Gunflint Lake on the west. Like the much earlier Fig Tree series, the Gunflint formation has been the subject of detailed geological investigation.

Granite underlies the Gunflint formation unconformably, that is, the basement rock had been eroded before the Gunflint sediments were deposited on its surface. Radioactive clocks have yielded two independent dates for the granite. A concentrate of its biotite contents indicates an age, in terms of the argon-potassium ratio, of 2.5 ± .75 billion years. The age of a whole-rock sample, in terms of its rubidium-strontium ratio, is 2.36 ± .70 billion years. The age of the granite thus provides a "floor" of maximum age under the Gunflint formation. The Gunflint cherts cannot be any older than these dates.

Micas separated from rocks in the upper levels of the Gunflint formation, collected near Thunder Bay, indicate an age in terms of the argon-potassium ratio of 1.60 ± .05 billion years. For technical reasons this figure is only 80 percent of the true age; it must therefore be adjusted to 1.90 ± .20 billion years. The micas thus provide a "ceiling" of minimum age for the cherts that lie at the base of the Gunflint formation. It seems reasonable to set the age of the cherts at approximately two billion years, which makes them a billion years or so younger than the Fig Tree cherts.

The only rocks in the Gunflint formation that contain microfossils are the cherts. Like the Fig Tree cherts, they are evidently the product of deposition in an aqueous environment that was rich in silica. Most of the Gunflint fossils are three-dimensional, and many show ex-

quisite anatomical details. It has been argued that the structure of these fossils has been preserved by the infiltration of silica from the surrounding sediments. In my opinion the organisms were preserved without distortion by being deposited in a siliceous solution that later crystallized into chert, much as a modern biological specimen is preserved by being embedded in plastic. The soft structure of the organism owes its preservation to the almost complete incompressibility of the silica matrix. This is as unusual as it is fortunate. In most instances of fossil preservation the matrix is composed of relatively plastic sediments that are much compressed during consolidation, with the result that any preservation of soft tissues, let alone preservation in three-dimensional form, is a rarity.

How were the Gunflint cherts deposited? The picture in Ontario is rather clearer than it is in the Transvaal. The Gunflint cherts were apparently precipitated and consolidated around an underlying complex of basement rocks, consisting of greenstone boulders and finer conglomerates, that seems to have been continuously submerged at the time. "Domes" of algae, which are visible to the unaided eye in samples of Gunflint rock, grew on the surface of the boulders. Algal "pillars" grew perpendicularly to the domes; their fossil remains consist of alternate layers of coarsely crystalline quartz and fine-grained black chert [*see bottom illustration on opposite page*].

In the 1950's I was privileged to be associated with Stanley A. Tyler of the University of Wisconsin in collecting and analyzing specimens of Gunflint chert from the Schreiber area. Only a few preliminary studies of these and other Gunflint specimens were published before Tyler died, although by then we both knew that his work had added to the fossil record an entire new group of very ancient, primitive photosynthetic organisms whose existence had not been suspected. Indeed, even though many years of study have now been devoted to the Gunflint organisms, the formation continues to yield new finds. Eight genera of primitive Gunflint plants, comprising 12 species, have been described so far, yet accompanying this article are illustrations of new forms of undetermined taxonomic status.

The most abundant of the Gunflint microfossils are filamentous structures. The majority of them are between .6 micron and 1.6 microns in diameter but a few are more than five microns across. They vary in length up to several hundred microns. Some of the filaments have inter-

SAMPLE OF GUNFLINT CHERT of Ontario (*above*) has a knobby surface. The knobs, exposed by weathering, are tops of pillars formed by algae in shallow water where Gunflint organisms lived.

VERTICAL SECTION through a chert sample (*below*) shows the structure of the algal pillars. Layers of quartz and of fossil-bearing black chert alternate in each pillar like sets of nesting thimbles.

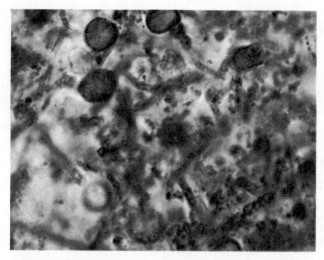

DENSE MIXTURE of organic detritus, spherical bodies and filaments is enlarged 250 diameters in this photomicrograph. All the micrographs on this page show thin sections of the Gunflint cherts.

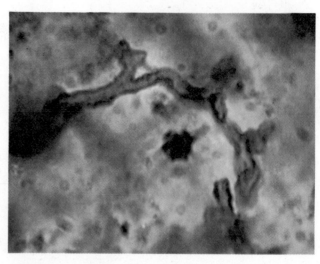

TUBULAR FILAMENT with branches and swellings is not obviously related to any living organism. Where it is not swollen it is about two microns across. It is called *Archaeorestis schreiberensis*.

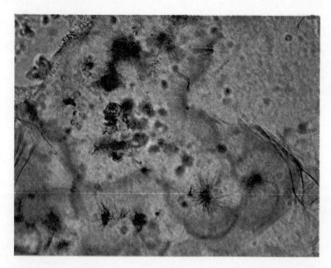

COLONY OF ALGAE is enlarged 200 diameters. The organisms, each a cluster of spikelets, have been named *Paleorivularia ontarica* because of their resemblance to the living genus *Rivularia*.

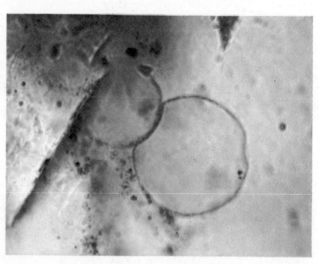

CLOSE CONTACT between a larger and a smaller cell is apparent in a chert section enlarged 1,000 diameters. Interrelations of this kind may represent a stage in the evolution of the eukaryotic cell.

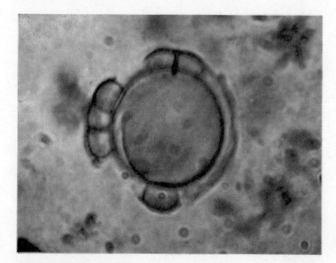

SPHERE WITHIN SPHERE, the inner one held away from the outer one by a number of smaller spheroids, is about 30 microns in diameter. The spheres comprise the extinct species *Eosphaera tyleri*.

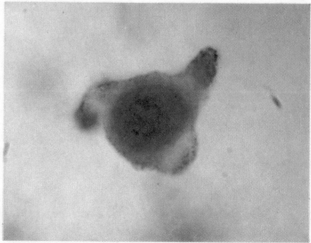

ASYMMETRICAL CELL, found in a recently prepared chert section, has not yet been formally classified. The Gunflint cherts have been studied for nearly 20 years, but they still yield new organisms.

nal walls at right angles to their length; others lack such walls. Where the walls are present, they can be broad or narrow. These primitive plants have been assigned on the basis of their morphology to four genera including five species. Among the present-day blue-green algae they resemble are the filamentous *Oscillatoria* and related forms; one of the four also shows resemblances to the modern iron-oxidizing bacterium *Crenothrix*.

Another genus of Gunflint organisms is represented by small spheroids, ranging from one micron in diameter to more than 16 microns. The walls of the spheroids vary both in thickness and in structural detail. Because of these variations they have been assigned to three separate species and placed in the genus *Huroniospora*. The spherical organisms are found in chert collected from many parts of the Gunflint formation but they are not equally abundant everywhere.

What kinds of organism are represented by the spheroids? We have only their simple morphology to judge by. Like their counterparts in the much older Fig Tree cherts, they might be noncolonial blue-green algae of the coccoid group. They might also, however, be the reproductive spores produced by the filamentous plants mentioned above, and some might even be the spores of iron bacteria. Still another possibility is that they are the fossilized bodies of free-swimming organisms whose flagella have not been preserved. Further study may lead to a choice among these alternatives.

The remaining Gunflint genera that have been intensively studied and described so far all show unusual characteristics. The organisms in one of the three genera are star-shaped and made up of radially arrayed filaments. The diameter of the "star" ranges from eight to 25 microns; in rare cases some of the filaments are branched. Although representatives of the genus are few in number and often poorly preserved, they are found throughout the Gunflint cherts.

The genus has been named *Eoastrion*—"dawn star"—to indicate the great antiquity and distinctive shape of its members. Two species have been established, one to accommodate the fossils with nonbranched filaments and the other the fossils with branched filaments. There are no clear analogies between the fossils and modern organisms, although in some respects *Eoastrion* resembles the curious iron- and manganese-oxidizing organism *Metallogenium personatum*.

A most peculiar organism comprises the next of these genera. Its fossils are

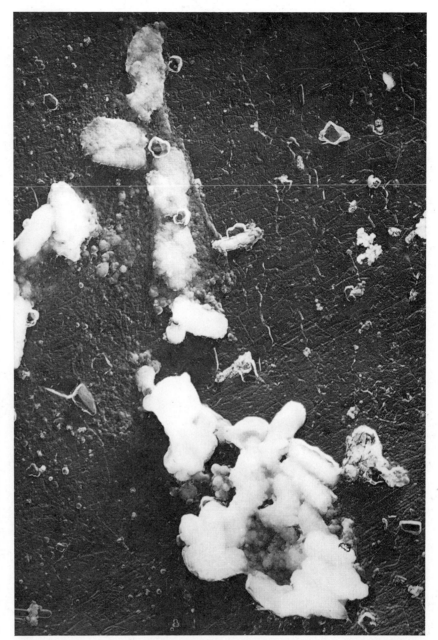

FOSSIL BACTERIA found in the Gunflint cherts are enlarged 30,000 diameters in this electron micrograph. Some two billion years old, they look like living rod-shaped species.

abundant in the cherts exposed near Kakabeka Falls, some 20 miles west of Thunder Bay. It consists of a bulb with a narrow stalk, surmounted by a structure that in some cases resembles an umbrella. The relative size of the three parts varies widely from specimen to specimen; the size of the bulb and the stalk together or of the umbrella alone ranges between 10 and 30 microns.

This odd plant has been assigned to the new genus *Kakabekia* and the species *umbellata* (which refers to the umbrella). Living organisms that superficially resemble it in form are certain multicellular polyps. All modern polyps, however, are many times larger.

The story of *Kakabekia* has had some interesting and continuing overtones. Late in 1964, quite independently (in fact, unaware) of paleontological investigations of the Gunflint fossils, Sanford M. Siegel discovered a strange new form while he was studying soil microorganisms that can survive extreme atmospheric conditions. The organism defied identification or assignment to any known taxonomic category. Siegel set it aside as being an enigma, although he kept his preparations, drawings and photomicrographs. A few months later a description of *Kakabekia* was published, and Siegel immediately noted a striking resemblance between his soil organism

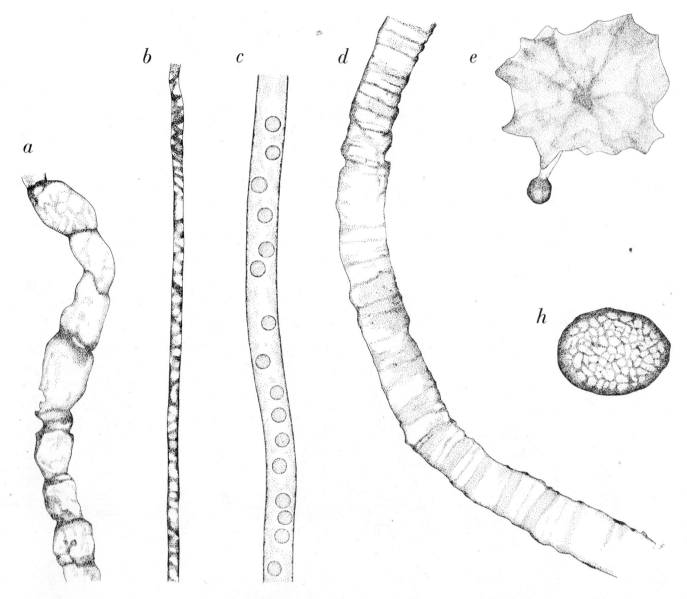

a   *b*   *c*   *d*   *e*

h

DIVERSITY OF FORM, evident among the plants fossilized in the Gunflint cherts, is indicative of evolutionary progress during the billion years that separate Fig Tree from Gunflint times. These composite drawings are based on a study of several specimens of each species and include three of the most unusual ones found. All are shown at the same scale: 2,500 times actual size. Filamentous organisms include two of the genus *Gunflintia*: *G. grandis* (*a*) and *G. minuta* (*b*). The other two are *Entosphaeroides amplus* (*c*) and *Animikiea septata* (*d*). The hydra-like *Kakabekia umbellata* (*e*) has a modern counterpart, if not a descendant, in a newfound soil

and some of the Gunflint specimens. Siegel's organism is very slow-growing, contains no chlorophyll and apparently has no nucleus; it may therefore be representative of a new group of prokaryotic microorganisms. It was first found in ammonia-rich soil collected at Harlech Castle in Wales, and it has since been recognized in soils from Alaska and Iceland and recently in soil from the slopes of the volcano Haleakala in Hawaii. Whether or not it is related to the two-billion-year-old *Kakabekia* is questionable, but the existence of two such bizarre forms is at least a remarkable evolutionary coincidence.

The eighth genus of Gunflint organisms comes from a single area near the easternmost outcropping of the chert, in the vicinity of Schreiber Beach. The organism consists of two concentric spheres; its outside diameter ranges from 28 to 32 microns. In most of the fossils the inner sphere is kept from contact with the outer one by as many as a dozen "spacers" in the form of small flattened spheroids. I have assigned this distinctive organism to the new genus *Eosphaera* ("dawn sphere") and the species *tyleri* (in honor of Tyler). No analogous living organism is known, nor has an organism resembling *Eosphaera* been found in any other Precambrian rocks. *Eosphaera* may be somewhat fancifully regarded as a "mistake" in evolution that did not survive the middle Precambrian.

Anyone who wanted to question that the organisms preserved in the Fig Tree cherts three billion years ago were photosynthetic could defend the negative position quite eloquently. Where the billion-years-younger Gunflint organisms are concerned, however, the affirmative evidence seems overwhelming. For one thing, the chemical analysis of organic material from the Gunflint cherts in several laboratories reveals the presence of the hydrocarbons pristane and phytane: two "chemical fossils" that can most reasonably be regarded as being breakdown products of chlorophyll. A second datum is the striking resemblance between the filamentous fossils that are the most abundant Gunflint forms and modern

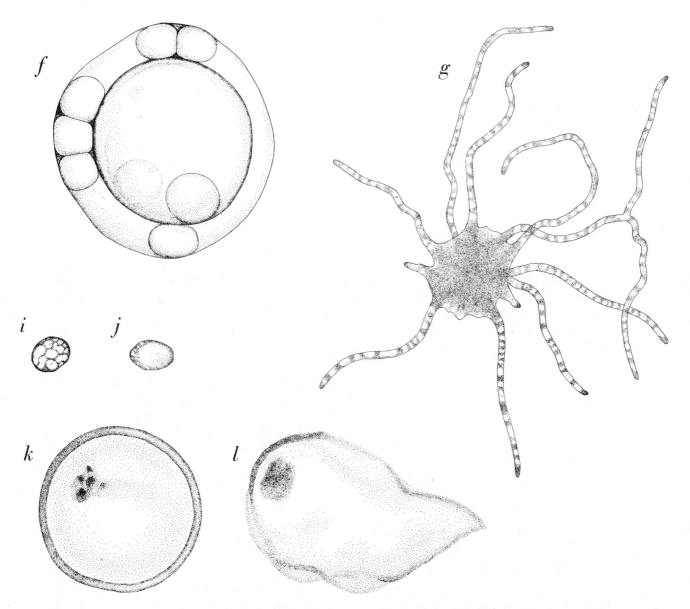

organism. The globular *Eosphaera tyleri* (*f*) evidently failed to survive Precambrian times. *Eoastrion bifurcatum* (*g*), with its array of filaments, is one of two such species. Three spherical organisms that differ chiefly in surface markings are assigned to the single genus *Huroniospora*; they are *H. microreticulata* (*h*), *H. macroreticulata* (*i*) and *H. psilata* (*j*). A one-celled organism that has not yet been classified (*k and l*) has internal features suggestive of a nucleus; the second specimen may have been fossilized as the process of cell division began. If these are eukaryotic cells, a key evolutionary event took place far earlier than is now thought.

photosynthetic blue-green algae. A third is the small domes and pillars in the Gunflint chert that resemble the structures formed in shallow water today by dome-building photosynthetic algae.

One may add the evidence of the relative abundance in Gunflint organic matter of the two nonradioactive carbon isotopes carbon 12 and carbon 13. The carbon in the carbon dioxide of the earth's atmosphere normally consists of about 99 percent carbon 12 and 1 percent carbon 13. In the process of photosynthesis, however, plants tend to fix slightly more carbon 12 than carbon 13, so that plant tissues are even poorer in the heavier isotope. Measurements of the ratio of the two isotopes in Gunflint organic material were made by Thomas C. Hoering of the Geophysical Laboratory of the Carnegie Institution of Washington. The results indicate that the Gunflint material too is poor in carbon 13, and to a degree that is almost identical with the depletion in modern algae and other photosynthetic plants. The billion-years-older Fig Tree organic material shows a carbon-12-to-carbon-13 ratio that is about the same as the ratio in the Gunflint material. This result lends credence to the view that the alga-like Fig Tree organisms probably were photosynthetic too.

It seems reasonable to conclude that even if oxygen was only a trivial component of the environment in Fig Tree times, the Gunflint organisms represent the intermediate agency that brought about the oxygen-rich environment at the end of the Precambrian. This is a development of major importance. It is not, however, the only development or even the most important development of Gunflint times. The variety of form (and therefore presumably of function) represented by the eight genera of Gunflint plants demonstrates that terrestrial life had crossed the second evolutionary threshold—the threshold of diversification—no less than two billion years ago.

The crossing of the third threshold is clearly documented in a series of late Precambrian formations, consisting primarily of limestones, sandstones and dolomites, found along the northern rim of the Amadeus basin in the Northern Territory of Australia. One member of the series is the Bitter Springs formation; a ridge in the Ross River area, about 40 miles from Alice Springs, consists of strata from its lower and middle levels. The exposed formations include isolated beds of dense black chert and laminated rocks that are associated with fossil structures built by colonial algae.

No absolute age is known for the Bitter Springs formation. Its top strata, however, lie some 4,000 feet below the lowest of the rocks that in the Ross River area are the boundary between formations of Precambrian age and those of the succeeding Cambrian period. The generally accepted date for the beginning of the Cambrian period is 600 million years ago, so that the lower position of the Bitter Springs cherts makes them considerably older. The Bitter Springs formation also underlies Precambrian sediments that are known to be some 820 million years old on the basis of their rubidium-strontium ratio. I think it is reasonable to assume that the Bitter Springs cherts are roughly a billion years old. This makes them only half as old as the Gunflint cherts and less than a third as old as the Fig Tree cherts. I collected samples of the Bitter Springs cherts in the Ross River area in April, 1965, and I added them to the Fig Tree samples for study with Schopf that summer.

A preliminary analysis of the abundant microfossils in the Bitter Springs cherts indicates that at least four general groups of lower plants lived in the shallow seas or embayments that covered this part of central Australia in late Precambrian times. As one would anticipate, the plants included filamentous blue-green algae akin to such living forms as *Oscillatoria* and *Nostoc*. Some of these filaments are more than 75 microns long; they are thickest (about 1.4 microns) at the center and taper toward the ends to less than one micron.

Our most exciting identifications during the preliminary analysis concerned the other three groups. On the basis of the internal structures that have been preserved, all three appear to represent various green algae. The green algae, unlike the blue-green algae, are eukaryotic! Thus the Bitter Springs cherts apparently contain the earliest-known fossil evidence that an organism potentially capable of sexual reproduction, or at least possessing a nucleus, had finally evolved.

By the time Schopf had finished his detailed analysis of the Bitter Springs specimens in 1968, he had concluded that they represent a total of three bacterium-like species, some 20 certain or probable representatives of blue-green algae, two certain genera of green algae, two possible species of fungi and two problematical forms. For one of the green algae, *Glenobotrydion aenigmatis*, an unusual coincidence of fossilization has preserved several specimens at various stages of mitotic division. By arranging the specimens in order one can re-create almost the entire sequence [*see illustration on page 78*].

With the discovery of a late Precambrian population of eukaryotic plants we approach the end of our story. The unequivocal appearance of the eukaryotic level of cellular organization at this point in earth history provides a welcome explanation for one of the principal puzzles of terrestrial evolution: Why do so many groups of higher organisms not appear in the fossil record until the span of geologic time is seven-eighths past? It is a good question, particularly in the light of what we know about the multicellular animals and higher algae early in the Paleozoic era.

Until recently the commonest explanation of the rarity of advanced organisms, multicellular animals in particular, before Cambrian times was that the supply of free oxygen in both the atmosphere and the hydrosphere had up to that point been extremely limited. Time was required to let the first prokaryotic organisms adjust to existence in an environment that had begun to contain this highly reactive element. Time was also needed for enough oxygen to leave the hydrosphere (where it was being produced by algal photosynthesis) and enter the atmosphere to establish a protective shield of ozone (O_3) between the earth's surface and the sun's harsh ultraviolet radiation. Until this shield existed neither shallow water nor bare ground would have been habitable.

An oxygen-poor environment was certainly a major obstacle in the path of

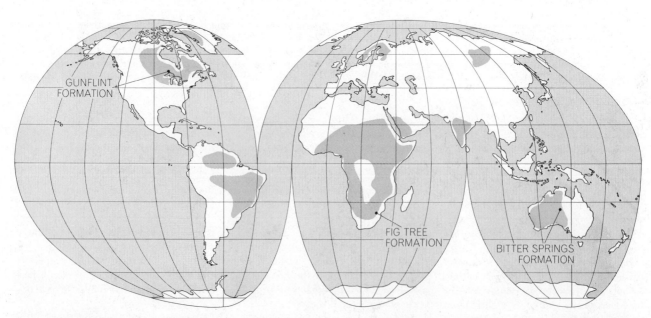

PRECAMBRIAN ROCKS occupy the old, low-lying "core" areas of continents throughout the world. Many, formed from accumulated sediments, are so altered by heat and pressure that any fossils have been obliterated. Some relatively unaltered exposures of Precambrian sediments, however, are rich in organic remains. Three such fossil-bearing Precambrian formations are shown on this map.

oxygen-dependent heterotrophs. In addition to the evidence provided by oxidized Precambrian formations, however, a number of recent studies indicate that the earth's atmosphere had actually accumulated enough oxygen to establish some kind of ozone shield considerably before the end of the Precambrian. The appearance of eukaryotic organisms late in the Precambrian, as indicated by the green algae of the Bitter Springs formation, provides a better explanation for the failure of higher organisms to appear until an even later time. The fundamental key to evolutionary progress is genetic variability. Sexual reproduction, which involves the recombination of heritable characteristics, is the highway to genetic variability and all its consequences, including the increased complexity of form and function at all levels of organization, that are thereafter apparent in the course of evolution.

This last of three thresholds, which separates the primitive world of cells without a nucleus from a world where sexual reproduction is possible, must have been crossed a long time before the formation of the Bitter Springs cherts. Only half a billion years or so later the Paleozoic seas were swarming with highly differentiated aquatic plants and animals, evolved from primitive forebears that had managed to cross all three Precambrian thresholds. Half a billion years does not seem to be evolutionary "room" enough to account for such epic progress. Moreover, there is evidence to suggest that developments in this direction may have begun in Gunflint times.

The Fig Tree, Gunflint and Bitter Springs cherts are not the only sources of Precambrian fossils, nor have Tyler, Schopf and I been the only investigators of the Precambrian fossil record. Indeed, work in the field has gone on for nearly a century. The emphasis has changed in recent years from an earlier concern with the curious "boundary" that has been traditionally accepted as separating the Precambrian from the beginning of the Paleozoic. We and our colleagues in many countries are primarily interested today in evidence that we hope will reveal even finer details of fossil cellular organization. It is a field that requires all levels of observation, from the macroscopic down to the electron microscopic. Workers in the field agree that the search for other stages and thresholds in the evolution of life must be focused on fine structures wherever these have been fortuitously preserved. The Precambrian fossil record is still meager, but significant gaps in the record are steadily

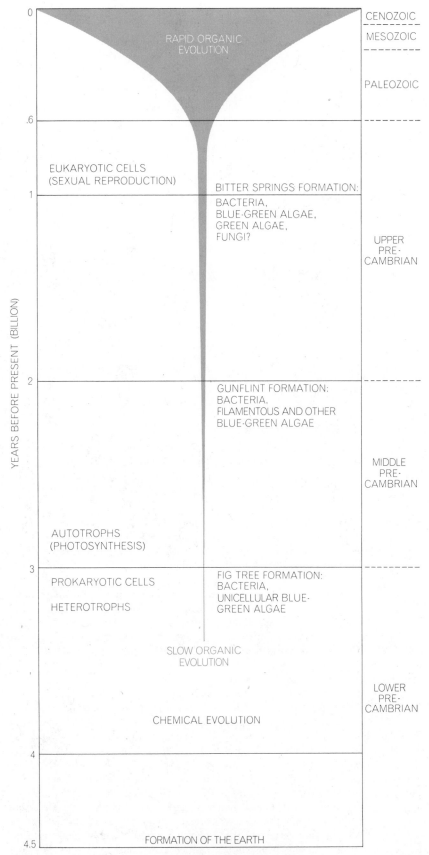

ORGANIC EVOLUTION is presented in terms of successively briefer episodes of biological advance. The Precambrian interval, the earliest and by far the longest, began when the earth was formed 4.5 billion years ago and ended 600 million years ago with the beginning of the Paleozoic era. The increasing abundance of species with the passage of time is indicated in color. Once organisms with eukaryotic, or truly nucleated, cells evolved, hastening evolutionary progress in late Precambrian times, the number of species multiplied explosively.

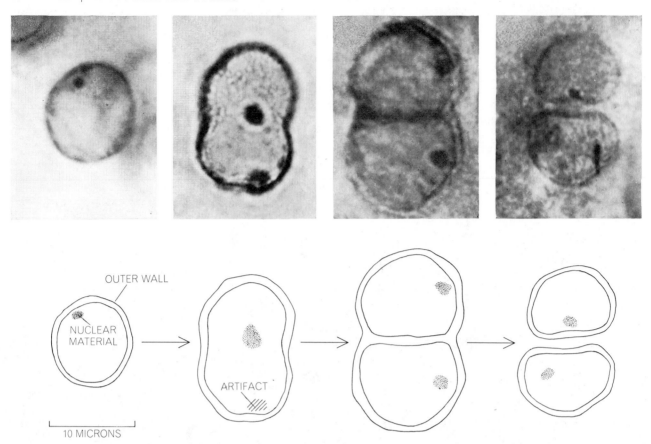

CELL DIVISION of a eukaryotic organism, a green alga of the genus *Glenobotrydion*, appears in the billion-year-old cherts of the Bitter Springs formation of Australia. J. William Schopf reconstructed the event (*drawings*), working with fossils that preserved individual organisms at different stages of division (*micrographs*). Existence of such biologically advanced algae in Bitter Springs times is proof that the evolutionary threshold leading to sexual reproduction and genetic variability had been crossed even earlier.

being filled. Indeed, the traditional concept of a clearly delimited Precambrian boundary may soon disappear from reconstructions of the history of terrestrial life.

Two final comments may be of interest to those who, like myself, are stimulated by the search for first causes no matter how often the search goes unrewarded. Some biologists now suggest that the organelles characteristic of the eukaryotic cell may once have been independent organisms that somehow came to live symbiotically inside larger host cells. It is not clear whether or not one of the host cell's responses to the presence of such a guest was a regrouping into a nucleus of the genetic material formerly scattered throughout the host's cytoplasm. If symbiosis was indeed the first step toward evolution of the eukaryotic cell, it may be that certain of the Gunflint organisms currently being studied show initial steps in this direction [*see middle illustration at right on page 50*]. A fascinating account of this transformation of various bacterial and algal organisms into components of the eukaryotic cell is given by Lynn Margulis of Boston University in a recent book. Mrs. Margulis' argument rests on biological grounds; the increasingly detailed Precambrian fossil record supports her thesis.

Astronomers and physicists have found evidence in recent years that molecules such as the hydroxyl radical (OH), carbon monoxide (CO), ammonia (NH_3), hydrogen cyanide (HCN) and formaldehyde (HCHO) are formed in the "empty" reaches of space. So far the ultimate in interstellar chemical complexity is represented by organic material that some investigators maintain they find in a peculiar class of meteorites known as carbonaceous chondrites. Until recently no carbonaceous chondrite had been recovered under circumstances that completely ruled out the possibility of accidental contamination by terrestrial organic material. As a result nagging questions about the chondrites have remained, particularly with regard to such biologically important molecules as amino acids (which are a *sine qua non* of terrestrial life). Recently, however, using stringent laboratory procedures to analyze carefully documented samples of the Murchison chondrite (which fell in southern Australia in September, 1969), Ponnamperuma and his associates at the Ames Research Center (in collaboration with Carleton B. Moore of Arizona State University and Ian R. Kaplan of the University of California at Los Angeles) seem to have proved the existence of extraterrestrial amino acids. Not only is the quantity of amino acids in the Murchison meteorite surprisingly high but also certain of them are unknown in terrestrial organisms and hence cannot be contaminants from the soil.

This finding opens up a new world of chemical evolution: a world of random synthetic processes not on the earth but in space, including the extraterrestrial formation of bodies (of which the carbonaceous chondrites seem to be fragments) that are rich in organic materials. The finding brings us back to the discussion at the beginning of this article. The chemical evolution of organic matter, the prelude to biogenesis on the earth, seems to have occurred elsewhere in the solar system or outside it. For knowledge of the earliest stages of organic evolution the biologist and paleontologist must rely on the chemist and astrophysicist.

Chemical Fossils

by Geoffrey Eglinton and Melvin Calvin
January 1967

Certain rocks as much as three billion years old have been found to contain organic compounds. What these compounds are and how they may have originated in living matter is under active study

If you ask a child to draw a dinosaur, the chances are that he will produce a recognizable picture of such a creature. His familiarity with an animal that lived 150 million years ago can of course be traced to the intensive studies of paleontologists, who have been able to reconstruct the skeletons of extinct animals from fossilized bones preserved in ancient sediments. Recent chemical research now shows that minute quantities of organic compounds—remnants of the original carbon-containing chemical constituents of the soft parts of the animal—are still present in some fossils and in ancient sediments of all ages, including some measured in billions of years. As a result of this finding organic chemists and geologists have joined in a search for "chemical fossils": organic molecules that have survived unchanged or little altered from their original structure, when they were part of organisms long since vanished.

This kind of search does not require the presence of the usual kind of fossil—a shape or an actual hard form in the rock. The fossil molecules can be extracted and identified even when the organism has completely disintegrated and the organic molecules have diffused into the surrounding material. In fact, the term "biological marker" is now being applied to organic substances that show pronounced resistance to chemical change and whose molecular structure gives a strong indication that they could have been created in significant amounts only by biological processes.

One might liken such resistant compounds to the hard parts of organisms that ordinarily persist after the soft parts have decayed. For example, hydrocarbons, the compounds consisting only of carbon and hydrogen, are comparatively resistant to chemical and biological attack. Unfortunately many other biologi-

cally important molecules such as nucleic acids, proteins and polysaccharides contain many bonds that hydrolyze, or cleave, readily; hence these molecules rapidly decompose after an organism dies. Nevertheless, several groups of workers have reported finding constituents of proteins (amino acids and peptide chains) and even proteins themselves in special well-protected sites, such as between the thin sheets of crystal in fossil shells and bones [see "Paleobiochemistry," by Philip H. Abelson; SCIENTIFIC AMERICAN, Offprint 101].

Where complete destruction of the organism has taken place one cannot, of course, visualize its original shape from the nature of the chemical fossils it has left behind. One may, however, be able to infer the biological class, or perhaps even the species, of organism that gave rise to them. At present such deductions must be extremely tentative because they involve considerable uncertainty. Although the chemistry of living organisms is known in broad outline, biochemists even today have identified the principal constituents of only a few small groups of living things. Studies in comparative biochemistry or chemotaxonomy are thus an essential parallel to organic geochemistry. A second uncertainty involves the question of whether or not the biochemistry of ancient organisms was generally the same as the biochemistry of present-day organisms. Finally, little is known of the chemical changes wrought in organic substances when they are entombed for long periods of time in rock or a fossil matrix.

In our work at the University of California at Berkeley and at the University of Glasgow we have gone on the assumption that the best approach to the study of chemical fossils is to analyze geological materials that have had a relatively simple biological and geological history.

The search for suitable sediments requires a close collaboration between the geologist and the chemist. The results obtained so far augur well for the future.

Organic chemistry made its first major impact on the earth sciences in 1936, when the German chemist Alfred Treibs isolated metal-containing porphyrins from numerous crude oils and shales. Certain porphyrins are important biological pigments; two of the best-known are chlorophyll, the green pigment of plants, and heme, the red pigment of the blood. Treibs deduced that the oils were biological in origin and could not have been subjected to high temperatures, since that would have decomposed some of the porphyrins in them. It is only during the past decade, however, that techniques have been available for the rapid isolation and identification of organic substances present in small amounts in oils and ancient sediments. Further refinements and new methods will be required for detailed study of the tiny amounts of organic substances found in some rocks. The effort should be worthwhile, because such techniques for the detection and definition of the specific architecture of organic molecules should not only tell us much more about the origin of life on the earth but also help us to establish whether or not life has developed on other planets. Furthermore, chemical fossils present the organic chemist with a new range of organic compounds to study and may offer the geologist a new tool for determining the environment of the earth in various geological epochs and the conditions subsequently experienced by the sediments laid down in those epochs.

If one could obtain the fossil molecules from a single species of organism, one would be able to make a direct correlation between present-day biochemistry

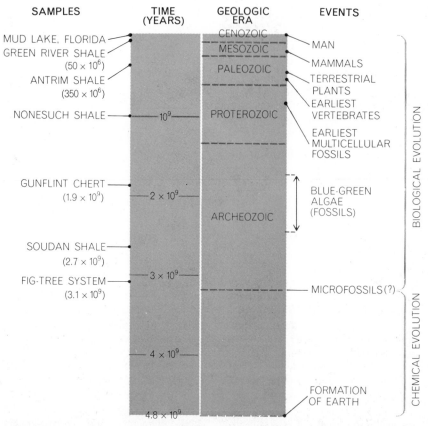

GEOLOGICAL TIME SCALE shows the age of some intensively studied sedimentary rocks (*left*) and the sequence of major steps in the evolution of life (*right*). The stage for biological evolution was set by chemical evolution, but the period of transition is not known.

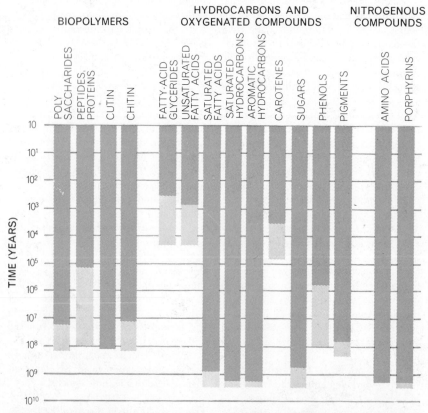

ORGANIC COMPOUNDS originally synthesized by living organisms and more or less modified have now been found in many ancient rocks that began as sediments. The dark bars indicate reasonably reliable identification; the light bars, unconfirmed reports. Cutin and chitin are substances present respectively in the outer structures of plants and of insects.

and organic geochemistry. For example, one could directly compare the lipids, or fatty compounds, isolated from a living organism with the lipids of its fossil ancestor. Unfortunately the fossil lipids and other fossil compounds found in sediments almost always represent the chemical debris from many organisms.

The deposition of a compressible fine-grained sediment containing mineral particles and disseminated organic matter takes place in an aquatic environment in which the organic content can be partially preserved; an example would be the bottom of a lake or a delta. The organic matter makes up something less than 1 percent of many ancient sediments. The small portion of this carbon-containing material that is soluble in organic solvents represents a part of the original lipid content, more or less modified, of the organisms that lived and died while the sediment was being deposited.

The organic content presumably consists of varying proportions of the components of organisms—terrestrial as well as aquatic—that have undergone chemical transformation while the sediment was being laid down and compressed. Typical transformations are reduction, which has the effect of removing oxygen from molecules and adding hydrogen, and decarboxylation, which removes the carboxyl radical (COOH). In addition, it appears that a variety of reactive unsaturated compounds (compounds having available chemical bonds) combine to form an insoluble amorphous material known as kerogen. Other chemical changes that occur with the passage of time are related to the temperature to which the rock is heated by geologic processes. Thus many petroleum chemists and geologists believe petroleum is created by progressive degradation, brought about by heat, of the organic matter that is finely disseminated throughout the original sediment. The organic matter that comes closest in structure to the chains and rings of carbon atoms found in the hydrocarbons of petroleum is the matter present in the lipid fraction of organisms. Another potential source of petroleum hydrocarbons is kerogen itself, presumably formed from a wide variety of organic molecules; it gives off a range of straight-chain, branched-chain and ring-containing hydrocarbons when it is strongly heated in the laboratory. One would also like to know more about the role of bacteria in the early steps of sediment formation. In the upper layers of most newly formed sediments there is strong bacterial activity, which must surely result in extensive alteration of the initially deposited organic matter.

In this article we shall concentrate on the isolation of fossil hydrocarbons. The methods must be capable of dealing with the tiny quantity of material available in most rocks. Our general procedure is as follows.

After cutting off the outer surface of a rock specimen to remove gross contaminants, we clean the remaining block with solvents and pulverize it. We then place the powder in solvents such as benzene and methanol to extract the organic material. Before this step we sometimes dissolve the silicate and carbonate minerals of the rock with hydrofluoric and hydrochloric acids. We separate the organic extract so obtained into acidic, basic and neutral fractions. The compounds in these fractions are converted, when necessary, into derivatives that make them suitable for separation by the technique of chromatography. For the initial separations we use column chromatography, in which a sample in solution is passed through a column packed with alumina or silica. Depending on their nature, compounds in the sample pass through the column at different speeds and can be collected in fractions as they emerge.

In subsequent stages of the analysis finer fractionations are achieved by means of gas-liquid chromatography. In this variation of the technique, the sample is vaporized into a stream of light gas, usually helium, and brought in contact with a liquid that tends to trap the compounds in the sample in varying degree. The liquid can be supported on an inorganic powder, such as diatomaceous earth, or coated on the inside of a capillary tube. Since the compounds are alternately trapped in the liquid medium and released by the passing stream of gas they progress through the column at varying speeds, with the result that they are separated into distinct fractions as they emerge from the tube. The temperature of the column is raised steadily as the separation proceeds, in order to drive off the more strongly trapped compounds.

The initial chromatographic separation is adjusted to produce fractions that consist of a single class of compound, for example the class of saturated hydrocarbons known as alkanes. Alkane molecules may consist either of straight chains of carbon atoms or of chains that include branches and rings. These subclasses can be separated with the help of molecular sieves: inorganic substances, commonly alumino-silicates, that have a fine honeycomb structure. We use a sieve whose

mesh is about five angstrom units, or about a thousandth of the wavelength of green light. Straight-chain alkanes, which resemble smooth flexible rods about 4.5 angstroms in diameter, can enter the sieve and are trapped. Chains with branches and rings are too big to enter and so are held back. The straight-chain alkanes can be liberated from the sieve for further analysis by dissolving the sieve in hydrofluoric acid. Other families of molecules can be trapped in special crystalline forms of urea and thiourea, whose crystal lattices provide cavities with diameters of five angstroms and seven angstroms respectively.

The families of molecules isolated in this way are again passed through gas

chromatographic columns that separate the molecular species within the family. For example, a typical chromatogram of straight-chain alkanes will show that molecules of increasing chain length emerge from the column in a regularly spaced sequence that parallels their increasing boiling points, thus producing a series of evenly spaced peaks. Although the species of molecule in a particular peak can often be identified tentatively on the basis of the peak's position, a more precise identification is usually desired. To obtain it one must collect the tiny amount of substance that produced the peak—often measured in micrograms—and examine it by one or more analytical methods such as ultraviolet and infrared

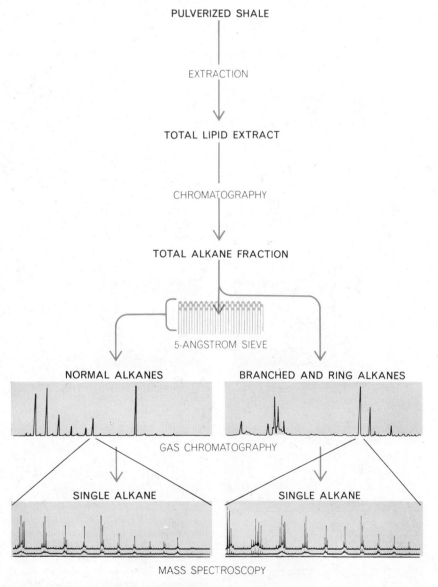

ANALYTICAL PROCEDURE for identifying chemical fossils begins with the extraction of alkane hydrocarbons from a sample of pulverized shale. In normal alkanes the carbon atoms are arranged in a straight chain. (Typical alkanes are illustrated on page 83.) Molecular sieves are used to separate straight-chain alkanes from alkanes with branched chains and rings. The two broad classes are then further fractionated. Compounds responsible for individual peaks in the chromatogram are identified by mass spectrometry and other methods.

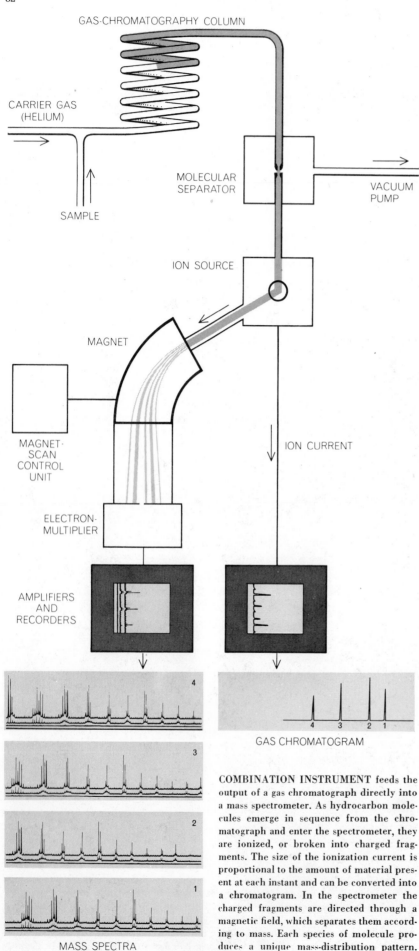

GAS-CHROMATOGRAPHY COLUMN

CARRIER GAS (HELIUM)

SAMPLE

MOLECULAR SEPARATOR

VACUUM PUMP

ION SOURCE

MAGNET

MAGNET-SCAN CONTROL UNIT

ION CURRENT

ELECTRON-MULTIPLIER

AMPLIFIERS AND RECORDERS

GAS CHROMATOGRAM

MASS SPECTRA

COMBINATION INSTRUMENT feeds the output of a gas chromatograph directly into a mass spectrometer. As hydrocarbon molecules emerge in sequence from the chromatograph and enter the spectrometer, they are ionized, or broken into charged fragments. The size of the ionization current is proportional to the amount of material present at each instant and can be converted into a chromatogram. In the spectrometer the charged fragments are directed through a magnetic field, which separates them according to mass. Each species of molecule produces a unique mass-distribution pattern.

spectroscopy, mass spectrometry or nuclear magnetic resonance. In one case X-ray crystallography is being used to arrive at the structure of a fossil molecule.

A new and useful apparatus is one that combines gas chromatography and mass spectrometry [*see illustration at right*]. The separated components emerge from the chromatograph and pass directly into the ionizing chamber of the mass spectrometer, where they are broken into submolecular fragments whose abundance distribution with respect to mass is unique for each component. These various analytical procedures enable us to establish a precise structure and relative concentration for each organic compound that can be extracted from a sample of rock.

How is it that such comparatively simple substances as the alkanes should be worthy of geochemical study? There are several good reasons. Alkanes are generally prominent components of the soluble lipid fraction of sediments. They survive geologic time and geologic conditions well because the carbon-hydrogen and carbon-carbon bonds are strong and resist reaction with water. In addition, alkane molecules can provide more information than the simplicity of their constitution might suggest; even a relatively small number of carbon and hydrogen atoms can be joined in a large number of ways. For example, a saturated hydrocarbon consisting of 19 carbon atoms and 40 hydrogen atoms could exist in some 100,000 different structural forms that are not readily interconvertible. Analysis of ancient sediments has already shown that in some cases they contain alkanes clearly related to the long-chain carbon compounds of the lipids in present-day organisms [*see illustration on next page*]. Generally one finds a series of compounds of similar structure, such as the normal, or straight-chain, alkanes (called *n*-alkanes); the compounds extracted from sediments usually contain up to 35 carbon atoms. Alkanes isolated from sediments may have been buried as such or formed by the reduction of substances containing oxygen.

The more complicated the structure of the molecule, the more valuable it is likely to be for geochemical purposes: its information content is greater. Good examples are the alkanes with branches and rings, such as phytane and cholestane. It is unlikely that these complex alkanes could be built up from small subunits by processes other than biological ones, at least in the proportions found. Hence we are encouraged to look for biological precursors with appropriate preexisting carbon skeletons.

In conducting this kind of search one makes the assumption, at least at the outset, that the overall biochemistry of past organisms was similar to that of present-day organisms. When lipid fractions are isolated directly from modern biological sources, they are generally found to contain a range of hydrocarbons, fatty acids, alcohols, esters and so on. The mixture is diverse but by no means random. The molecules present in such fractions have structures that reflect the chemical reaction pathways systematically followed in biological organisms. There are only a few types of biological molecule wherein long chains of carbon atoms are linked together; two examples are the straight-chain lipids, the end groups of which may include oxygen atoms, and the lipids known as isoprenoids.

The straight-chain lipids are produced by what is called the polyacetate pathway [*see illustration on opposite page*]. This pathway leads to a series of fatty acids with an even number of carbon atoms; the odd-numbered molecules are missing. One also finds in nature straight-chain alcohols (*n*-alkanols) that likewise have an even number of carbon atoms, which is to be expected if they are formed by simple reduction of the corresponding fatty acids. In contrast, the straight-chain hydrocarbons (*n*-alkanes) contain an *odd* number of carbon atoms. Such a series would be produced by the decarboxylation of the fatty acids.

The second type of lipid, the isoprenoids, have branched chains consisting of five-carbon units assembled in a regular order [*see illustration on page 85*]. Because these units are assembled in head-to-tail fashion the side-chain methyl groups (CH_3) are attached to every fifth carbon atom. (Tail-to-tail addition occurs less frequently but accounts for several important natural compounds, for example beta-carotene.) When the isoprenoid skeleton is found in a naturally occurring molecule, it is reasonable to assume that the compound has been formed by this particular biological pathway.

Chlorophyll is possibly the most widely distributed molecule with an isoprenoid chain; therefore it must make some contribution to the organic matter in sediments. Its fate under conditions of geological sedimentation is not known, but it may decompose into only two or three large fragments [*see illustration on page 88*]. The molecule of chlorophyll *a* consists of a system of intercon-

NORMAL-C_{29}

ISO-C_{18}

ANTEISO-C_{18}

CYCLOHEXYL-NORMAL-C_{17}

PHYTANE (C_{20} ISOPRENOID)

CHOLESTANE (C_{27} STERANE)

- HYDROGEN
- CARBON

GAMMACERANE (C_{30} TRITERPANE)

CAROTANE (C_{40} TETRATERPANE)

ALKANE HYDROCARBON MOLECULES can take various forms: straight chains (which are actually zigzag chains), branched chains and ring structures. Those depicted here have been found in crude oils and shales. The molecules shown in color are so closely related to well-known biological molecules that they are particularly useful in bespeaking the existence of ancient life. The broken lines indicate side chains that are directed into the page.

nected rings and a phytyl side chain, which is an isoprenoid. When chlorophyll is decomposed, it seems likely that the phytyl chain is split off and converted to phytane (which has the same number of carbon atoms) and pristane (which is shorter by one carbon atom). When both of these branched alkanes are found in a sediment, one has reasonable presumptive evidence that chlorophyll was once present. The chlorophyll ring system very likely gives rise to the metal-containing porphyrins that are found in many crude oils and sediments.

Phytane and pristane may actually enter the sediments directly. Max Blumer of the Woods Hole Oceanographic Institution showed in 1965 that certain species of animal plankton that eat the plant plankton containing chlorophyll store quite large quantities of pristane and related hydrocarbons. The animal plankton act in turn as a food supply for bigger marine animals, thereby accounting for the large quantities of pristane in the liver of the shark and other fishes.

An indirect source for the isoprenoid alkanes could be the lipids found in the outer membrane of certain bacteria that live only in strong salt solutions, an environment that might be found where ancient seas were evaporating. Morris Kates of the National Research Council of Canada has shown that a phytyl-containing lipid (diphytyl phospholipid) is common to bacteria with the highest salt requirement but not to the other bacteria examined so far.

This last example brings out the point that in spite of the overall oneness of present-day biochemistry, organisms do differ in the compounds they make. They also synthesize the same compounds in different proportions. These differences are making it possible to classify living species on a chemotaxonomic, or chemical, basis rather than on a morphological, or shape, basis. Eventually it may be possible to extend chemical classification to ancient organisms, creating a discipline that could be called paleochemotaxonomy.

Our study of chemical fossils began in 1961, when we decided to probe the sedimentary rocks of the Precambrian period in a search for the earliest signs of life. This vast period of time, some four billion years, encompasses the beginnings of life on this planet and its early development to the stage of organisms consisting of more than one cell [see illustrations on page 80]. We hoped that our study would complement the efforts being made by a number of work-

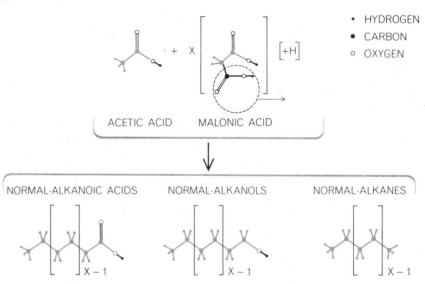

- HYDROGEN
- CARBON
- OXYGEN

STRAIGHT-CHAIN LIPIDS are created in living organisms from simple two-carbon and three-carbon compounds: acetate and malonate, shown here as their acids. The complex biological process, which involves coenzyme A, is depicted schematically. The fatty acids (n-alkanoic acids) and fatty alcohols (n-alkanols) produced in this way have an even number of carbon atoms. The removal of carbon dioxide from the fatty acids, the net effect of decarboxylation, would give rise to a series of n-alkanes with an odd number of carbon atoms.

ers, including one of us (Calvin), to imitate in the laboratory the chemical evolution that must have preceded the appearance of life on earth. We also saw the possibility that our work could be adapted to the study of meteorites and of rocks obtained from the moon or nearby planets. Thus it even includes the possibility of uncovering exotic and alien biochemistries. The exploration of the ancient rocks of the earth provides a testing ground for the method and the concepts involved.

We chose the alkanes because one might expect them to resist fairly high temperatures and chemical attack for long periods of time. Moreover, J. G. Bendoraitis of the Socony Oil Company, Warren G. Meinschein of the Esso Research Laboratory and others had already identified individual long-chain alkanes, including a range of isoprenoid types, in certain crude oils. Even more encouraging, J. J. Cummins and W. E. Robinson of the U.S. Bureau of Mines had just made a preliminary announcement of their isolation of phytane, pristane and other isoprenoids from a relatively young sedimentary rock: the Green River shale of Colorado, Utah and Wyoming. Thus the alkanes seemed to offer the biological markers we were seeking. Robinson generously provided our laboratory with samples of the Green River shale, which was deposited some 50 million years ago and constitutes the major oil-shale reserve of the U.S.

The Green River shale, which is the

remains of large Eocene lakes in a rather stable environment, contains a considerable fraction (.6 percent) of alkanes. Using the molecular-sieve technique, we split the total alkane fraction into alkanes with straight chains and those with branched chains and rings and ran the resulting fractions through the gas chromatograph [see illustration at top left on page 86]. The straight-chain alkanes exhibit a marked dominance of molecules containing an odd number of carbon atoms, which is to be expected for straight-chain hydrocarbons from a biological source. The other fraction shows a series of prominent sharp peaks; we conclusively identified them as isoprenoids, confirming the results of Cummins and Robinson. The large proportion of phytane, the hydrocarbon corresponding to the entire side chain of chlorophyll, is particularly noteworthy. The oxygenated counterparts of the steranes and triterpanes (27 to 30 carbon atoms) and the high-molecular-weight n-alkanes (29 to 31 carbons) are typical constituents of the waxy covering of the leaves and pollen of land plants, leading to the inference that such plants made major contributions to the organic matter deposited in the Green River sediments.

Although the gross chemical structure (number of rings and side chains) of the steranes and triterpanes was established in this work, it was only recently that the precise structure of one of these hydrocarbons was conclusively established. E. V. Whitehead and his associates in the British Petroleum Company and Robinson and his collaborators in the Bureau of Mines have shown that one of the triterpanes extracted from the Green River shale is identical in all respects with gammacerane [*see illustration on page 83*]. Conceivably it is produced by the reduction of a compound known as gammaceran-3-beta-ol, which was recently isolated from the common protozoon *Tetrahymena pyriformis*. Other derivatives of gammacerane are rather widely distributed in the plant kingdom.

At our laboratory in Glasgow, Sister Mary T. J. Murphy and Andrew McCormick recently identified several steranes and triterpanes and also the tetraterpane called perhydro-beta-carotene, or carotane [*see top illustration on page 90*]. Presumably carotane is derived by reduction from beta-carotene, an important red pigment of plants. A similar reduction process could convert the familiar biological compound cholesterol into cholestane, one of the steranes found in the Green River shale [*see same illustration on page 90*]. The mechanism and sedimentary site of such geochemical reduction processes is an important problem awaiting attack.

W. H. Bradley of the U.S. Geological Survey has sought a contemporary counterpart of the richly organic ooze that presumably gave rise to the Green River shale. So far he has located only four lakes, two in the U.S. and two in Africa, that seem to be reasonable candidates. One of them, Mud Lake in Flor-

ida, is now being studied closely. A dense belt of vegetation surrounding the lake filters out all the sand and silt that might otherwise be washed into it from the land. As a result the main source of sedimentary material is the prolific growth of microscopic algae. The lake bottom uniformly consists of a grayish-green ooze about three feet deep. The bottom of the ooze was deposited about 2,300 years ago, according to dating by the carbon-14 technique.

Microscopic examination of the ooze shows that it consists mainly of minute fecal pellets, made up almost exclusively of the cell walls of blue-green algae. Some pollen grains are also present. Decay is surprisingly slow in spite of the ooze's high content of bound oxygen and the temperatures characteristic of Florida. Chemical analyses in several laboratories, reported this past November at a meeting of the Geological Society of America, indicate that there is indeed considerable correspondence between the lipids of the Mud Lake ooze and those of the Green River shale. Eugene McCarthy of the University of California at Berkeley has also found beta-carotene in samples of Mud Lake ooze that are about 1,100 years old. The high oxygen content of the Mud Lake ooze seems inconsistent, however, with the dominance of oxygen-poor compounds in the Green River shale. The long-term geological mechanisms that account for the loss of oxygen may have to be sought in sediments older than those in Mud Lake.

Sediments much older than the Green River shale have now been examined by our groups in Berkeley and Glasgow, and by workers in other universities and in oil-industry laboratories. We find that the hydrocarbon fractions in these more ancient samples are usually more complex than those of the Green River shale; the gas chromatograms of the older samples tend to show a number of partially resolved peaks centered around a single maximum. One of the older sediments we have studied is the Antrim shale of Michigan. A black shale probably 350 million years old, it resembles other shales of the Chattanooga type that underlie many thousands of square miles of the eastern U.S. Unlike the Green River shale, the straight-chain alkane fraction of the Antrim shale shows little or no predominance of an odd number of carbon atoms over an even number [*see middle illustration of three at top of pages 86 and 87*]. The alkanes with branched chains and rings, however, continue to be rich in isoprenoids.

The fact that alkanes with an odd number of carbon atoms are not predominant in the Antrim shale and sediments of comparable antiquity may be owing to the slow cracking by heat of carbon chains both in the alkane component and in the kerogen component. The effect can be partially reproduced in the laboratory by heating a sample of the Green River shale for many hours above 300 degrees centigrade. After such treatment the straight-chain alkanes show a reduced dominance of odd-carbon molecules and the branched-chain-and-ring fraction is more complex.

The billion-year-old shale from the Nonesuch formation at White Pine, Mich., exemplifies how geological, geochemical and micropaleontological techniques can be brought to bear on the problem of detecting ancient life. With the aid of the electron microscope Elso S. Barghoorn and J. William Schopf of Harvard University have detected in the Nonesuch shale "disaggregated particles of condensed spheroidal organic matter." In collaboration with Meinschein the Harvard workers have also found evidence that the Nonesuch shale contains isoprenoid alkanes, steranes and porphyrins. Independently we have analyzed the Nonesuch shale and found that it contains pristane and phytane, in addition to iso-alkanes, anteiso-alkanes and cyclohexyl alkanes.

Barghoorn and S. A. Tyler have also detected microfossils in the Gunflint chert of Ontario, which is 1.9 billion years old, almost twice the age of the

ACETIC ACID (3 UNITS) MEVALONIC ACID ISOPENTENYL PYROPHOSPHATE

DIMER, $(C_5)_2$

TRIMER, $(C_5)_3$

TETRAMER, $(C_5)_4$

POLYMER $(C_5)_n$

BRANCHED-CHAIN LIPIDS are produced in living organisms by an enzymatically controlled process, also depicted schematically. In this process three acetate units link up to form a six-carbon compound (mevalonic acid), which subsequently loses a carbon atom and is combined with a high-energy phosphate. "Head to tail" assembly of the five-carbon subunits produces branched-chain molecules that are referred to as isoprenoid structures.

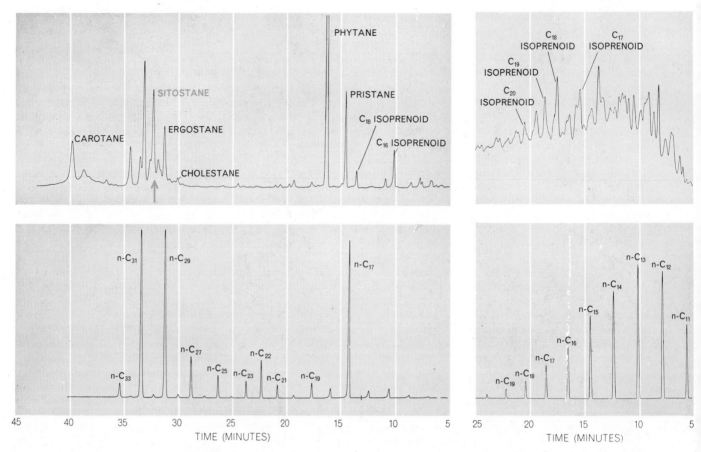

HYDROCARBONS IN YOUNG SEDIMENT, the 50-million-year-old Green River shale, produced these chromatograms. Alkanes with branched chains and rings appear in the top curve, normal alkanes in the bottom curve. The alkanes in individual peaks were identified by mass spectrometry and other methods. Such alkanes as phytane and pristane and the predominance of normal alkanes with an odd number of carbon atoms affirm that the hydrocarbons are biological in origin. The bimodal distribution of the curves is also significant.

OLDER SEDIMENTS are represented by the Antrim shale (*left*), which is 350 million years old, and by the Soudan shale (*right*), which is 2.7 billion years old. Alkanes with branched chains and rings again are shown in the top curves, normal alkanes

Nonesuch shale. They have reported that the morphology of the Gunflint microfossils "is similar to that of the existing primitive filamentous blue-green algae."

One of the oldest Precambrian sediments yet analyzed is the Soudan shale of Minnesota, which was formed about 2.7 billion years ago. Although its total

hydrocarbon content is only .05 percent, we have found that it contains a mixture of straight-chain alkanes and branched-chain-and-ring alkanes not unlike those present in the much younger Antrim shale [*see third illustration of three at top of next two pages*]. In the branched-chain-and-ring fraction we have identi-

fied pristane and phytane. Steranes and triterpanes also seem to be present, but we have not yet established their precise three-dimensional structure. Preston E. Cloud of the University of California at Los Angeles has reported that the Soudan shale contains microstructures resembling bacteria or blue-green algae,

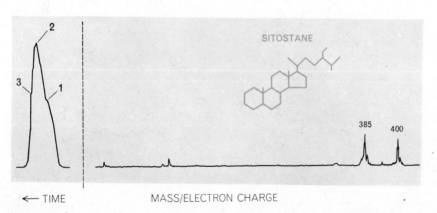

IDENTIFICATION OF SITOSTANE in the Green River shale was accomplished by "trapping" the alkanes that produced a major peak in the chromatogram (*colored arrow at top left on this page*) and passing them through the chromatograph-mass spectrometer. As the chromatograph drew the curve at the left, three scans were made with the spectrometer. Scan 1 (*partially shown at right*) is identical with the scan produced by pure sitostane.

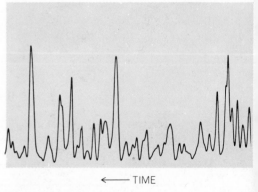

NINETEEN-CARBON ISOPRENOID was identified in the Antrim shale by using a co-injection technique together with a high-resolution gas chromatograph. These two high-resolution curves, each taken from a

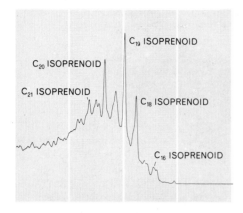

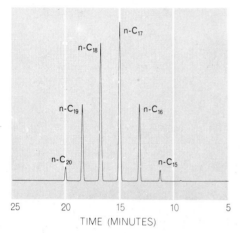

in the bottom curves. These chromatograms lack a pronounced bimodal distribution and the normal alkanes do not show a predominance of molecules with an odd number of carbon atoms. Nevertheless, the prevalence of isoprenoids argues for a biological origin.

but he is not satisfied that the evidence is conclusive.

A few reports are now available on the most ancient rocks yet examined: sediments from the Fig Tree system of Swaziland in Africa, some 3.1 billion years old. An appreciable fraction of the alkane component of these rocks consists

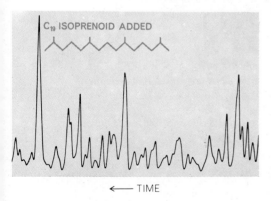

much longer trace, show the change in height of a specific peak when a small amount of pure 19-carbon isoprenoid was coinjected with the sample. Other peaks can be similarly identified by coinjecting known alkanes.

of isoprenoid molecules. If one assumes that isoprenoids are chemical vestiges of chlorophyll, one is obliged to conclude that living organisms appeared on the earth only about 1.7 billion years after the earth was formed (an estimated 4.8 billion years ago).

Before reaching this conclusion, however, one would like to be sure that the isoprenoids found in ancient sediments have the precise carbon skeleton of the biological molecules from which they are presumed to be derived. So far no sample of pristane or phytane—the isoprenoids that may be derived from the phytyl side chain of chlorophyll—has been shown to duplicate the precise three-dimensional structure of a pure reference sample. Vigorous efforts are being made to clinch the identification.

Assuming that one can firmly establish the presence of biologically structured isoprenoid alkanes in a sediment, further questions remain. The most serious one is: Were the hydrocarbons or their precursors deposited when the sediment was formed or did they seep in later? This question is not easily answered. A sample can be contaminated at any point up to—and after—the time it reaches the laboratory bench. Fossil fuels, lubricants and waxes are omnipresent, and laboratory solvents contain tiny amounts of pristane and phytane unless they are specially purified.

One way to determine whether or not rock hydrocarbons are indigenous is to measure the ratio of the isotopes carbon 13 and carbon 12 in the sample. (The ratio is expressed as the excess of carbon 13 in parts per thousand compared with the isotope ratio in a standard: a sample of a fossil animal known as a belemnite.) The principle behind the test is that photosynthetic organisms discriminate against carbon 13 in preference to carbon 12. Although we have few clues to the abundance of the two isotopes throughout the earth's history, we can at least test various hydrocarbon fractions in a given sample to see if they have the same isotope ratio. As a simple assumption, one would expect to find the same ratio in the soluble organic fraction as in the insoluble kerogen fraction, which could not have seeped into the rock as kerogen.

Philip H. Abelson and Thomas C. Hoering of the Carnegie Institution of Washington have made such measurements on sediments of various geological ages and have found that the isotope ratios for soluble and insoluble fractions in most samples agree reasonably well. In some of the oldest samples, however,

there are inconsistencies. In the Soudan shale, for example, the soluble hydrocarbons have an isotope ratio expressed as -25 parts per thousand compared with -34 parts per thousand for the kerogen. (In younger sediments and in present-day marine organisms the ratio is about midway between these two values: -29 parts per thousand.) The isotope divergence shown by hydrocarbons in the Soudan shale may indicate that the soluble hydrocarbons and the kerogen originated at different times. But since nothing is known of the mechanism of kerogen formation or of the alterations that take place in organic matter generally, the divergence cannot be regarded as unequivocal evidence of separate origin.

On the other hand, there is some reason to suspect that the isoprenoids did indeed seep into the Soudan shale sometime after the sediments had been laid down. The Soudan formation shows evidence of having been subjected to temperatures as high as 400 degrees C. The isoprenoid hydrocarbons pristane and phytane would not survive such conditions for very long. But since the exact date, extent and duration of the heating of the Soudan shale are not known, one can only speculate about whether the isoprenoids were indigenous and survived or seeped in later. In any event, they could not have seeped in much later because the sediment became compacted and relatively impervious within a few tens of millions of years.

A still more fundamental issue is whether or not isoprenoid molecules and others whose architecture follows that of known biological substances could have been formed by nonbiological processes. We and others are studying the kinds and concentrations of hydrocarbons produced by both biological and nonbiological sources. Isoprene itself, the hydrocarbon whose polymer constitutes natural rubber, is easily prepared in the laboratory, but no one has been able to demonstrate that isoprenoids can be formed nonbiologically under geologically plausible conditions. Using a computer approach, Margaret O. Dayhoff of the National Biomedical Research Foundation and Edward Anders of the University of Chicago and their colleagues have concluded that under certain restricted conditions isoprene should be one of the products of their hypothetical reactions. But this remains to be demonstrated in the laboratory.

It is well known, of course, that complex mixtures of straight-chain, branched-chain and even ring hydrocarbons can readily be synthesized in the

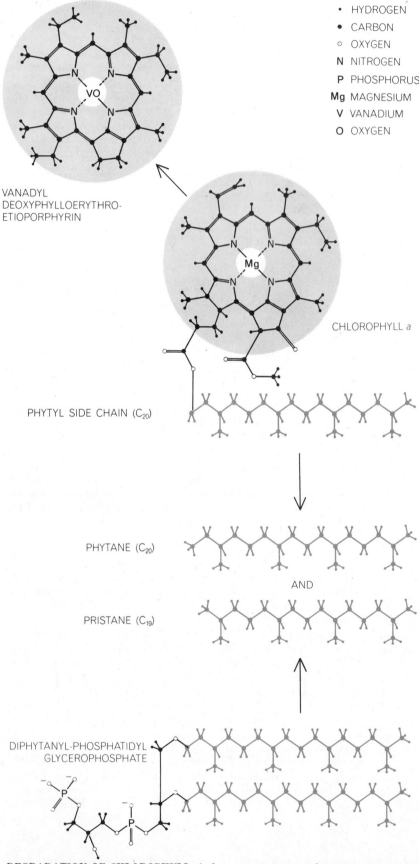

- • HYDROGEN
- • CARBON
- ○ OXYGEN
- N NITROGEN
- P PHOSPHORUS
- **Mg** MAGNESIUM
- V VANADIUM
- O OXYGEN

VANADYL
DEOXYPHYLLOERYTHRO-
ETIOPORPHYRIN

CHLOROPHYLL *a*

PHYTYL SIDE CHAIN (C$_{20}$)

PHYTANE (C$_{20}$)

AND

PRISTANE (C$_{19}$)

DIPHYTANYL-PHOSPHATIDYL
GLYCEROPHOSPHATE

DEGRADATION OF CHLOROPHYLL *A*, the green pigment in plants, may give rise to two kinds of isoprenoid molecules, phytane and pristane, that have been identified in many ancient sediments. It also seems likely that phytane and pristane can be derived from the isoprenoid side chains of a phosphate-containing lipid (*bottom structure*) that is a major constituent of salt-loving bacteria. The porphyrin ring of chlorophyll *a* is the probable source of vanadyl porphyrin (*upper left*) that is widely found in crude oils and shales.

laboratory from simple starting materials. For example, the Fischer-Tropsch process, used by the Germans as a source of synthetic fuel in World War II, produces a mixture of saturated hydrocarbons from carbon monoxide and water. The reaction requires a catalyst (usually nickel, cobalt or iron), a pressure of about 100 atmospheres and a temperature of from 200 to 350 degrees C. The hydrocarbons formed by this process, and several others that have been studied, generally show a smooth distribution of saturated hydrocarbons. Many of them have straight chains but lack the special characteristics (such as the predominance of chains with an odd number of carbons) found in the similar hydrocarbons present in many sediments. Isoprenoid alkanes, if they are formed at all, cannot be detected.

Paul C. Marx of the Aerospace Corporation has made the ingenious suggestion that isoprenoids may be produced by the hydrogenation of graphite. In the layered structure of graphite the carbon atoms are held in hexagonal arrays by carbon-carbon bonds. Marx has pointed out that if the bonds were broken in certain ways during hydrogenation, an isoprenoid structure might result. Again a laboratory demonstration is needed to support the hypothesis. What seems certain, however, is that nonbiological syntheses are extremely unlikely to produce those specific isoprenoid patterns found in the products of living cells.

Another dimension is added to this discussion by the proposal, made from time to time by geologists, that certain hydrocarbon deposits are nonbiological in origin. Two alleged examples of such a deposit are a mineral oil found enclosed in a quartz mineral at the Abbott mercury mine in California and a bitumen-like material called thucolite found in an ancient nonsedimentary rock in Ontario. Samples of both materials have been analyzed in our laboratory at Berkeley. The Abbott oil contains a significant isoprenoid fraction and probably constitutes an oil extracted and brought up from somewhat older sediments of normal biological origin. The thucolite consists chiefly of carbon from which only a tiny hydrocarbon fraction can be extracted. Our analysis shows, however, that the fraction contains trace amounts of pristane and phytane. Recognizing the hazards of contamination, we are repeating the analysis, but on the basis of our preliminary findings we suspect that the thucolite sample represents an oil of biological origin that has been almost completely carbonized. We are aware, of course, that one runs the risk of invoking circular arguments in such discussions. Do isoprenoids demonstrate biological origin (as we and others are suggesting) or does the presence of isoprenoids in such unlikely substances indicate that they were formed nonbiologically? The debate may not be quickly settled.

There is little doubt, in any case, that organic compounds of considerable variety and complexity must have accumulated on the primitive earth during the prolonged period of nonbiological chemical development—the period of chemical evolution. With the appearance of the first living organisms biological evolution took command and presumably the "food stock" of nonbiological compounds was rapidly altered. If the changeover was abrupt on a geological time scale, one would expect to find evidence of it in the chemical composition of sediments whose age happens to bracket the period of transition. Such a discontinuity would make an intensely exciting find for organic geochemistry. The transition from chemical to biological evolution must have occurred earlier than three billion years ago. As yet, however, no criteria have been established for distinguishing between the two types of evolutionary process.

We suggest that an important distinction should exist between the kinds of molecules formed by the two processes. In the period of chemical evolution autocatalysis must have been one of the dominant mechanisms for creating large molecules. An autocatalytic system is one in which a particular substance promotes the formation of more of itself. In biological evolution, on the other hand, two different molecular systems are involved: an information-bearing system based on nucleic acids and a catalytic system based on proteins. The former directs the synthesis of the latter. A major problem, subject to laboratory experiment, is visualizing how the two systems originated and were linked.

The role of lipids in the transition may have been important. Today lipids form an important part of the membranes of all living cells. A. I. Oparin, the Russian investigator who was among the first to discuss in detail the chemical origin of life, has suggested that an essential step in the transition from chemical to biological evolution may have been the formation of membranes around droplets, which could then serve as "reaction vessels." Such self-assembling membranes might well have required lipid constituents for their function, which would be to allow some compounds to enter and leave the "cell"

more readily than others. These membranes might have been formed nonbiologically by the polymerization of simple two-carbon and three-carbon units. According to this line of reasoning, the compounds that are now prominent constituents of living things are prominent precisely because they were prominent products of chemical evolution. We scarcely need add that this is a controversial and therefore stimulating hypothesis.

What one can say with some confidence is that autocatalysis alone seems unlikely to have been capable of producing the distribution pattern of hydrocarbons observed in ancient Precambrian rocks, even when some allowance is made for subsequent reactions over the course of geologic time. That it could have produced compounds of the observed type is undoubtedly possible, but

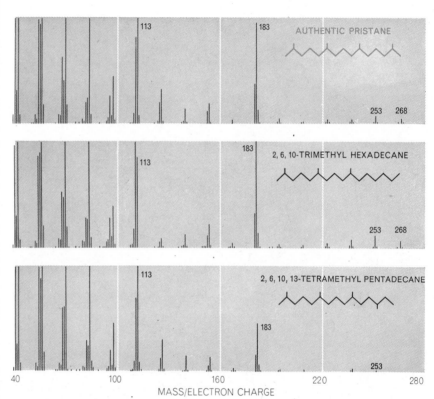

SIMILARITY OF MASS SPECTRA makes it difficult to distinguish the 19-carbon isoprenoid pristane from two of its many isomers (molecules with the same number of carbon and hydrogen atoms). The three records shown here are replotted from the actual tracings produced by pure compounds. When the sample contains impurities, as is normally the case, the difficulty of identifying authentic pristane by mass spectrometry is even greater.

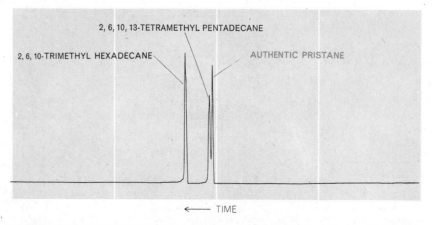

IDENTIFICATION OF PRISTANE can be done more successfully with the aid of a high-resolution gas chromatograph. When pure pristane and the isomers shown in the illustration above are fed into such an instrument, they produce three distinct peaks. This curve and the mass spectra were made by Eugene McCarthy of the University of California at Berkeley. He also made the isoprenoid study shown at the bottom of pages 86 and 87.

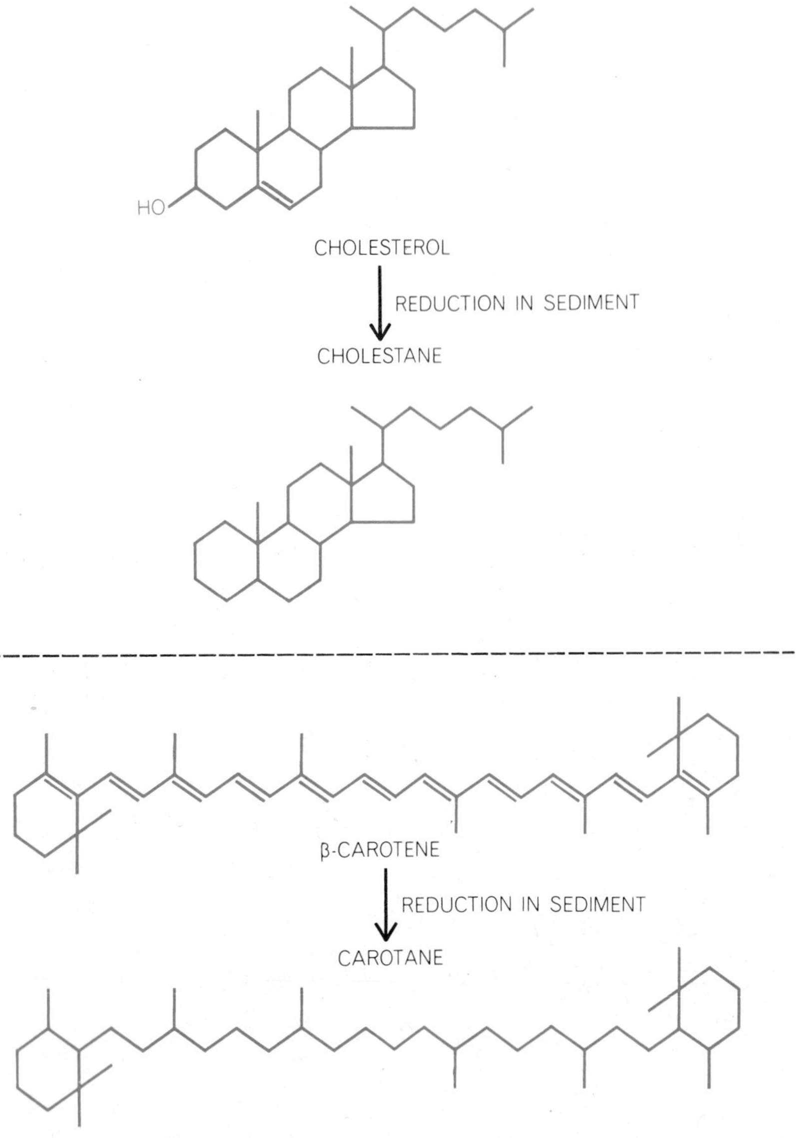

TWO ALKANES IN GREEN RIVER SHALE, cholestane and carotane, probably have been derived from two well-known biological substances: cholesterol and beta-carotene. The former is closely related to the steroid hormones; the latter is a red pigment widely distributed in plants. These two natural substances can be converted to their alkane form by reduction: a process that adds hydrogen at the site of double bonds and removes oxygen.

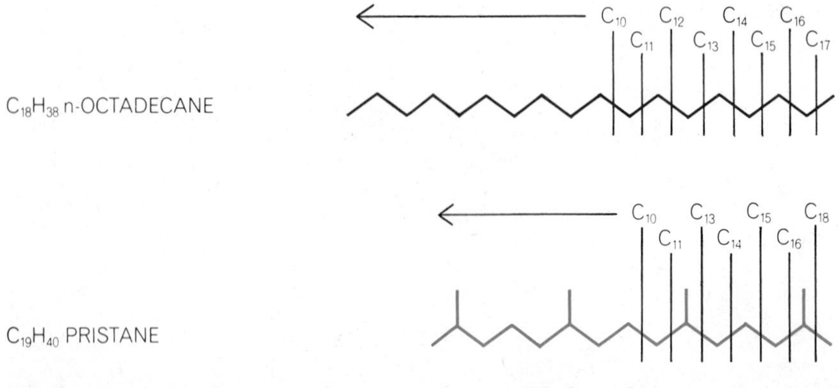

$C_{18}H_{38}$ n-OCTADECANE

$C_{19}H_{40}$ PRISTANE

EFFECT OF HEATING ALKANES is to produce a smoothly descending series of products (normal alkenes) if the starting material is a straight-chain molecule such as n-octadecane. (The term "alkene" denotes a hydrocarbon with one carbon-carbon double bond.) If, however, the starting material is an isoprenoid such as pristane, heating it to 600 degrees centigrade for .6 second produces an irregular series of alkenes because of the branched chain. Such degradation processes may take place in deeply buried sediments. These findings were made by R. T. Holman and his co-workers at the Hormel Institute in Minneapolis, Minn.

it seems to us that the observed pattern could not have arisen without the operation of those molecular systems we now recognize as the basis of living things. Eventually it should be possible to find in the geological record certain molecular fossils that will mark the boundary between chemical and biological evolution.

Another and more immediate goal for the organic geochemist is to attempt to trace on the molecular level the direction of biological evolution. For such a study one would like to have access to the actual nucleic acids and proteins synthesized by ancient organisms, but these are as yet unavailable (except perhaps in rare instances). We must therefore turn to the geochemically stable compounds, such as the hydrocarbons and oxygenated compounds that must have derived from the operation of the more perishable molecular systems. These "secondary metabolites," as we have referred to them, can be regarded as the signatures of the molecular systems that synthesized them or their close relatives.

It follows that the carbon skeletons found in the secondary metabolites of present-day organisms are the outcome of evolutionary selection. Thus it should be possible for the organic geochemist to arrange in a rough order of evolutionary sequence the carbon skeletons found in various sediments. There are some indications that this may be feasible. G. A. D. Haslewood of Guy's Hospital Medical School in London has proposed that the bile alcohols and bile acids found in present-day vertebrates can be arranged in an evolutionary sequence: the bile acids of the most primitive organisms contain molecules nearest chemically to cholesterol, their supposed biosynthetic precursor.

Within a few years the organic geochemist will be presented with a piece of the moon and asked to describe its organic contents. The results of this analysis will be awaited with immense curiosity. Will we find that the moon is a barren rock or will we discover traces of organic compounds—some perhaps resembling the complex carbon skeletons we had thought could be produced only by living systems? During the 1970's and 1980's we can expect to receive reports from robot sampling and analytical instruments landed on Mars, Venus and perhaps Jupiter. Whatever the results and their possible indications of alien forms of life, we shall be very eager to learn what carbon compounds are present elsewhere in the solar system.

IV

THE EARLY
EVOLUTION OF LIFE

IV

THE EARLY EVOLUTION OF LIFE

INTRODUCTION

We have seen that primitive cell-like structures can be found in the oldest rocks, and that 1000 m yr later oxygen-producing photosynthetic blue-green bacteria appeared. During this interval, protocells became truly biological by the evolution of a genetic apparatus. At present we can only speculate on the origin of the genetic code and the reason why certain base triplets are a universal code for the position of specific amino acids during protein synthesis. F. H. C. Crick describes clearly the nature of the genetic code and the protein-synthesizing apparatus in his remarkably concise article.

Once a genetic code arises, those cells having it become subject to Darwinian natural selection. This mechanism, which acts upon cells that are *almost* genetically identical and that reproduce to excess, serves to select those few that can reproduce faster. The biological time bomb goes off. Genetically endowed cells evolve to invade every conceivable facet of the environment. We are their descendants.

Lynn Margulis takes a careful look at present-day biology and presents a wealth of information showing how complex, eucaryotic animal and plant cells arose through symbiotic associations of simpler cells. A genetically endowed, bacterialike protocell stem gives rise to a variety of bacterial groups, including the oxygen-producing blue-green bacteria. Accumulation of atmospheric oxygen as a biosynthetic byproduct began about 2200 m yr ago. Free oxygen was toxic to most primitive stem cells, as it is to many groups of anaerobic bacteria today. But the availability of oxygen presented potential advantages too. Metabolism using oxygen—respiration—returns some 20 times more metabolic energy than fermentation, *if* the oxygen could be tolerated. All aerobic cells from bacteria to ourselves show a common pattern in their oxygen-detoxifying mechanisms.

Margulis suggests convincingly that free oxygen was the driving force in the evolution of the more complex eucaryotic cell from symbiotic associations of anaerobic bacterial stem cells. She also shows how eucaryotic organelles, such as mitochondria and chloroplasts, were probably once free bacterialike cells.

William Schopf, at the University of California, Los Angeles, has searched for microfossil evidence of the oldest eucaryotic cells, while they were still symbiotic. He found fossils with strong suggestions of organelles and nuclei; some fossils were unmistakably grouped in packets of four, reminiscent of meiotic cell division. Only eucaryotic cells have these characteristics. The oldest of these specimens is from 2200 m yr ago—a date that nicely matches Margulis's oxygen-driven theory of symbiosis.

Studies on present biology, microfossils, and the nature of the genetic apparatus all serve to elucidate events of great antiquity. Similarly, detailed studies of the molecular architecture of a single kind of protein can show how it evolves as a consequence of evolutionary pressures on the genetic/protein synthesizing apparatus.

Margaret Dayhoff compared the amino acid sequences of cytochrome c, a heme-containing protein vital in electron transport; it is found in some bacteria and in all mitochondria of all eucaryotic cells. Our ancestors more than 2500 m yr ago were synthesizing similar proteins. The crucial points concerning evolution are that family trees of relatedness can be constructed and that proteins also evolve by gene duplication.

Relatedness may be measured by the number of different amino acid changes present at each sequence position within proteins from different lines of descent. The greater the number of changes, the more distant on the evolutionary tree are the proteins—and hence the species being compared.

Fundamentally, this work extends and confirms the findings of studies of microfossils, biological cell structures, and the universality of the genetic code. All life has descended from a common stem cell population.

Suggested Further Reading

Gatlin, Lila L. 1972. *Information Theory and the Living System.* Columbia University Press, New York.

Gibor, Aharon (ed.). 1976. *Conditions for Life, Readings from Scientific American.* W. H. Freeman and Company, San Francisco.

Hood, Leroy E., John H. Wilson, and William B. Wood. 1974. *Molecular Biology of Eucaryotic Cells,* Vol. I. W. A. Benjamin, Menlo Park, Ca.

Margulis, Lyn. 1970. *Origin of the Eucaryotic Cell.* Yale University Press, New Haven.

Ohno, S. 1970. *Evolution by Gene Duplication.* Springer-Verlag, New York.

Rutten, M. G. 1971. *The Origin of Life by Natural Causes.* Elsevier, Amsterdam.

Srb, A. M., R. D. Owen, and R. S. Edgar (eds.). 1970. *Facets of Genetics, Readings from Scientific American,* Part I. W. H. Freeman and Company, San Francisco.

9   The Genetic Code: III

by F. H. C. Crick
October 1966

*The central theme of molecular biology is confirmed by
detailed knowledge of how the four-letter language
embodied in molecules of nucleic acid controls the
20-letter language of the proteins*

The hypothesis that the genes of the living cell contain all the information needed for the cell to reproduce itself is now more than 50 years old. Implicit in the hypothesis is the idea that the genes bear in coded form the detailed specifications for the

thousands of kinds of protein molecules the cell requires for its moment-to-moment existence: for extracting energy from molecules assimilated as food and for repairing itself as well as for replication. It is only within the past 15 years, however, that insight has been gained

into the chemical nature of the genetic material and how its molecular structure can embody coded instructions that can be "read" by the machinery in the cell responsible for synthesizing protein molecules. As the result of intensive work by many investigators the story

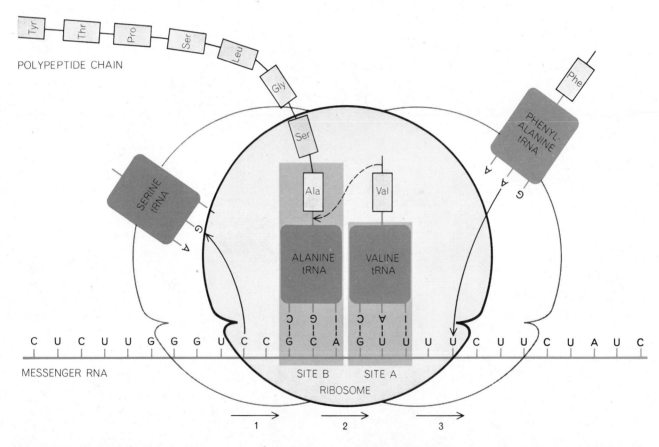

SYNTHESIS OF PROTEIN MOLECULES is accomplished by the intracellular particles called ribosomes. The coded instructions for making the protein molecule are carried to the ribosome by a form of ribonucleic acid (RNA) known as "messenger" RNA. The RNA code "letters" are four bases: uracil (U), cytosine (C), adenine (A) and guanine (G). A sequence of three bases, called a codon, is required to specify each of the 20 kinds of amino acid, identified here by their abbreviations. (A list of the 20 amino acids and their abbreviations appears on the next page.) When linked end to end, these amino acids form the polypeptide chains of which proteins are composed. Each type of amino acid is transported to the ribosome by a particular form of "transfer" RNA (tRNA), which carries an anticodon that can form a temporary bond with one of the codons in messenger RNA. Here the ribosome is shown moving along the chain of messenger RNA, "reading off" the codons in sequence. It appears that the ribosome has two binding sites for molecules of tRNA: one site (A) for positioning a newly arrived tRNA molecule and another (B) for holding the growing polypeptide chain.

of the genetic code is now essentially complete. One can trace the transmission of the coded message from its original site in the genetic material to the finished protein molecule.

The genetic material of the living cell is the chainlike molecule of deoxyribonucleic acid (DNA). The cells of many bacteria have only a single chain; the cells of mammals have dozens clustered together in chromosomes. The DNA molecules have a very long backbone made up of repeating groups of phosphate and a five-carbon sugar. To this backbone the side groups called bases are attached at regular intervals. There are four standard bases: adenine (A), guanine (G), thymine (T) and cytosine (C). They are the four "letters" used to spell out the genetic message. The exact sequence of bases along a length of the DNA molecule determines the structure of a particular protein molecule.

Proteins are synthesized from a standard set of 20 amino acids, uniform throughout nature, that are joined end to end to form the long polypeptide chains of protein molecules [see illustration at left]. Each protein has its own characteristic sequence of amino acids. The number of amino acids in a polypeptide chain ranges typically from 100 to 300 or more.

The genetic code is not the message itself but the "dictionary" used by the cell to translate from the four-letter language of nucleic acid to the 20-letter language of protein. The machinery of the cell can translate in one direction only: from nucleic acid to protein but not from protein to nucleic acid. In making this translation the cell employs a variety of accessory molecules and mechanisms. The message contained in DNA is first transcribed into the similar molecule called "messenger" ribonucleic acid—messenger RNA. (In many viruses—the tobacco mosaic virus, for example—the genetic material is simply RNA.) RNA too has four kinds of bases as side groups; three are identical with those found in DNA (adenine, guanine and cytosine) but the fourth is uracil (U) instead of thymine. In this first transcription of the genetic message the code letters A, G, T and C in DNA give rise respectively to U, C, A and G. In other words, wherever A appears in DNA, U appears in the RNA transcription; wherever G appears in DNA, C appears in the transcription, and so on. As it is usually presented the dictionary of the genetic code employs the letters found in RNA (U, C, A, G) rather than those found in DNA (A, G, T, C).

The genetic code could be broken

easily if one could determine both the amino acid sequence of a protein and the base sequence of the piece of nucleic acid that codes it. A simple comparison of the two sequences would yield the code. Unfortunately the determination of the base sequence of a long nucleic acid molecule is, for a variety of reasons, still extremely difficult. More indirect approaches must be used.

Most of the genetic code first became known early in 1965. Since then additional evidence has proved that almost all of it is correct, although a few features remain uncertain. This article describes how the code was discovered and some of the work that supports it.

Scientific American has already presented a number of articles on the genetic code. In one of them [" The Genetic Code," Offprint 123] I explained that the experimental evidence (mainly indirect) suggested that the code was a triplet code: that the bases on the messenger RNA were read three at a time and that each group corresponded to a particular amino acid. Such a group is called a codon. Using four symbols in groups of three, one can form 64 distinct triplets. The evidence indicated that most of these stood for one amino acid or another, implying that an amino acid was usually represented by several codons. Adjacent amino acids were coded by adjacent codons, which did not overlap.

In a sequel to that article ["The Genetic Code: II," Offprint 153]. Marshall W. Nirenberg of the National Institutes of Health explained how the composition of many of the 64 triplets had been determined by actual experiment. The technique was to synthesize polypeptide chains in a cell-free system, which was made by breaking open cells of the colon bacillus (*Escherichia coli*) and extracting from them the machinery ·for protein synthesis. Then the system was provided with an energy supply, 20 amino acids and one or another of several types of synthetic RNA. Although the exact sequence of bases in each type was random, the proportion of bases was known. It was found that each type of synthetic messenger RNA directed the incorporation of certain amino acids only.

By means of this method, used in a quantitative way, the *composition* of many of the codons was obtained, but the *order* of bases in any triplet could not be determined. Codons rich in G were difficult to study, and in addition a few mistakes crept in. Of the 40 codon compositions listed by Nirenberg

AMINO ACID	ABBREVIATION
ALANINE	Ala
ARGININE	Arg
ASPARAGINE	AspN
ASPARTIC ACID	Asp
CYSTEINE	Cys
GLUTAMIC ACID	Glu
GLUTAMINE	GluN
GLYCINE	Gly
HISTIDINE	His
ISOLEUCINE	Ileu
LEUCINE	Leu
LYSINE	Lys
METHIONINE	Met
PHENYLALANINE	Phe
PROLINE	Pro
SERINE	Ser
THREONINE	Thr
TRYPTOPHAN	Tryp
TYROSINE	Tyr
VALINE	Val

TWENTY AMINO ACIDS constitute the standard set found in all proteins. A few other amino acids occur infrequently in proteins but it is suspected in each case that they originate as one of the standard set and become chemically modified after they have been incorporated into a polypeptide chain.

in his article we now know that 35 were correct.

The Triplet Code

The main outlines of the genetic code were elucidated by another technique invented by Nirenberg and Philip Leder. In this method no protein synthesis occurs. Instead one triplet at a time is used to bind together parts of the machinery of protein synthesis.

Protein synthesis takes place on the comparatively large intracellular structures known as ribosomes. These bodies travel along the chain of messenger RNA, reading off its triplets one after another and synthesizing the polypeptide chain of the protein, starting at the amino end (NH_2). The amino acids do not diffuse to the ribosomes by themselves. Each amino acid is joined chemically by a special enzyme to one of the codon-recognizing molecules known both as soluble RNA (sRNA) and transfer RNA (tRNA). (I prefer the latter designation.) Each tRNA molecule has its own triplet of bases, called an anticodon, that recognizes the relevant codon on the messenger RNA by pairing bases with it [*see illustration on page 94*].

Leder and Nirenberg studied which amino acid, joined to its tRNA molecules, was bound to the ribosomes in the presence of a particular triplet, that is, by a "message" with just three letters. They did so by the neat trick of passing the mixture over a nitrocellulose filter that retained the ribosomes. All the tRNA molecules passed through the filter except the ones specifically bound to the ribosomes by the triplet. Which they were could easily be decided by using mixtures of amino acids in which one kind of amino acid had been made artificially radioactive, and determining the amount of radioactivity absorbed by the filter.

For example, the triplet GUU retained the tRNA for the amino acid valine, whereas the triplets UGU and UUG did not. (Here GUU actually stands for the trinucleoside diphosphate GpUpU.) Further experiments showed that UGU coded for cysteine and UUG for leucine.

Nirenberg and his colleagues synthesized all 64 triplets and tested them for their coding properties. Similar results have been obtained by H. Gobind Khorana and his co-workers at the University of Wisconsin. Various other groups have checked a smaller number of codon assignments.

Close to 50 of the 64 triplets give a clearly unambiguous answer in the binding test. Of the remainder some evince only weak binding and some bind more than one kind of amino acid. Other results I shall describe later suggest that the multiple binding is often an artifact of the binding method. In short, the binding test gives the meaning of the majority of the triplets but it does not firmly establish all of them.

The genetic code obtained in this way, with a few additions secured by other methods, is shown in the table below. The 64 possible triplets are set out in a regular array, following a plan

SECOND LETTER

		U	C	A	G	
U		UUU ⎫ Phe UUC ⎭ UUA ⎫ Leu UUG ⎭	UCU ⎫ UCC ⎪ Ser UCA ⎬ UCG ⎭	UAU ⎫ Tyr UAC ⎭ UAA OCHRE UAG AMBER	UGU ⎫ Cys UGC ⎭ UGA ? UGG Tryp	U C A G
C		CUU ⎫ CUC ⎪ Leu CUA ⎬ CUG ⎭	CCU ⎫ CCC ⎪ Pro CCA ⎬ CCG ⎭	CAU ⎫ His CAC ⎭ CAA ⎫ GluN CAG ⎭	CGU ⎫ CGC ⎪ Arg CGA ⎬ CGG ⎭	U C A G
A		AUU ⎫ AUC ⎬ Ileu AUA ⎭ AUG Met	ACU ⎫ ACC ⎪ Thr ACA ⎬ ACG ⎭	AAU ⎫ AspN AAC ⎭ AAA ⎫ Lys AAG ⎭	AGU ⎫ Ser AGC ⎭ AGA ⎫ Arg AGG ⎭	U C A G
G		GUU ⎫ GUC ⎪ Val GUA ⎬ GUG ⎭	GCU ⎫ GCC ⎪ Ala GCA ⎬ GCG ⎭	GAU ⎫ Asp GAC ⎭ GAA ⎫ Glu GAG ⎭	GGU ⎫ GGC ⎪ Gly GGA ⎬ GGG ⎭	U C A G

FIRST LETTER (left margin) · THIRD LETTER (right margin)

GENETIC CODE, consisting of 64 triplet combinations and their corresponding amino acids, is shown in its most likely version. The importance of the first two letters in each triplet is readily apparent. Some of the allocations are still not completely certain, particularly for organisms other than the colon bacillus (*Escherichia coli*). "Amber" and "ochre" are terms that referred originally to certain mutant strains of bacteria. They designate two triplets, UAA and UAG, that may act as signals for terminating polypeptide chains.

that clarifies the relations between them.

Inspection of the table will show that the triplets coding for the same amino acid are often rather similar. For example, all four of the triplets starting with the doublet AC code for threonine. This pattern also holds for seven of the other amino acids. In every case the triplets XYU and XYC code for the same amino acid, and in many cases XYA and XYG are the same (methionine and tryptophan may be exceptions). Thus an amino acid is largely selected by the first two bases of the triplet. Given that a triplet codes for, say, valine, we know that the first two bases are GU, whatever the third may be. This pattern is true for all but three of the amino acids. Leucine can start with UU or CU, serine with UC or AG and arginine with CG or AG. In all other cases the amino acid is uniquely related to the first two bases of the triplet. Of course, the converse is often not true. Given that a triplet starts with, say, CA, it may code for either histidine or glutamine.

Synthetic Messenger RNA's

Probably the most direct way to confirm the genetic code is to synthesize a messenger RNA molecule with a strictly defined base sequence and then find the amino acid sequence of the polypeptide produced under its influence. The most extensive work of this nature has been done by Khorana and his colleagues. By a brilliant combination of ordinary chemical synthesis and synthesis catalyzed by enzymes, they have made long RNA molecules with various repeating sequences of bases. As an example, one RNA molecule they have synthesized has the sequence UGUG-UGUGUGUG.... When the biochemical machinery reads this as triplets the message is UGU–GUG–UGU–GUG.... Thus we expect that a polypeptide will be produced with an alternating sequence of two amino acids. In fact, it was found that the product is Cys–Val–Cys–Val.... This evidence alone would not tell us which triplet goes with which amino acid, but given the results of the binding test one has no hesitation in concluding that UGU codes for cysteine and GUG for valine.

In the same way Khorana has made chains with repeating sequences of the type XYZ... and also XXYZ.... The type XYZ...would be expected to give a "homopolypeptide" containing one amino acid corresponding to the triplet XYZ. Because the starting point is not clearly defined, however, the homo-polypeptides corresponding to YZX... and ZXY... will also be produced. Thus poly-AUC makes polyisoleucine, polyserine and polyhistidine. This confirms that AUC codes for isoleucine, UCA for serine and CAU for histidine. A repeating sequence of four bases will yield a single type of polypeptide with a repeating sequence of four amino acids. The general patterns to be expected in each case are set forth in the table on this page. The results to date have amply demonstrated by a direct biochemical method that the code is indeed a triplet code.

Khorana and his colleagues have so far confirmed about 25 triplets by this method, including several that were quite doubtful on the basis of the binding test. They plan to synthesize other sequences, so that eventually most of the triplets will be checked in this way.

The Use of Mutations

The two methods described so far are open to the objection that since they do not involve intact cells there may be some danger of false results. This objection can be met by two other methods of checking the code in which the act of protein synthesis takes place inside the cell. Both involve the effects of genetic mutations on the amino acid sequence of a protein.

It is now known that small mutations are normally of two types: "base substitution" mutants and "phase shift" mutants. In the first type one base is changed into another base but the total number of bases remains the same. In the second, one or a small number of bases are added to the message or subtracted from it.

There are now extensive data on base-substitution mutants, mainly from studies of three rather convenient proteins: human hemoglobin, the protein of tobacco mosaic virus and the A protein of the enzyme tryptophan synthetase obtained from the colon bacillus. At least 36 abnormal types of human hemoglobin have now been investigated by many different workers. More than 40 mutant forms of the protein of the tobacco mosaic virus have been examined by Hans Wittmann of the Max Planck Institute for Molecular Genetics in Tübingen and by Akita Tsugita and Heinz Fraenkel-Conrat of the University of California at Berkeley [see "The Genetic Code of a Virus," by Heinz Fraenkel-Conrat; SCIENTIFIC AMERICAN Offprint 193]. Charles Yanofsky and his group at Stanford University have characterized about 25 different mutations of the A protein of tryptophan synthetase.

VARIETY OF SYNTHETIC RNA's with repeating sequences of bases have been produced by H. Gobind Khorana and his colleagues at the University of Wisconsin. They contain two or three different bases (X, Y, Z) in groups of two, three or four. When introduced into cell-free systems containing the machinery for protein synthesis, the base sequences are read off as triplets (middle) and yield the amino acid sequences indicated at the right.

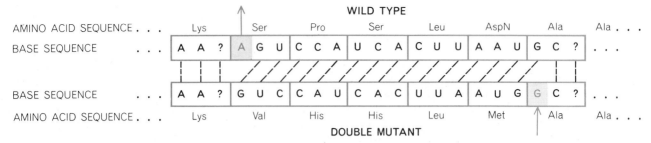

WILD TYPE

| AMINO ACID SEQUENCE... | Lys | Ser | Pro | Ser | Leu | AspN | Ala | Ala... |

"PHASE SHIFT" MUTATIONS help to establish the actual codons used by organisms in the synthesis of protein. The two partial amino acid sequences shown here were determined by George Streisinger and his colleagues at the University of Oregon. The sequences are from a protein, a type of lysozyme, produced by the bacterial virus T4. A pair of phase-shift mutations evidently removed one base, A, and inserted another, G, about 15 bases farther on. The base sequence was deduced theoretically from the genetic code.

The remarkable fact has emerged that in every case but one the genetic code shows that the change of an amino acid in a polypeptide chain could have been caused by the alteration of a single base in the relevant nucleic acid. For example, the first observed change of an amino acid by mutation (in the hemoglobin of a person suffering from sickle-cell anemia) was from glutamic acid to valine. From the genetic code dictionary on page 96 we see that this could have resulted from a mutation that changed either GAA to GUA or GAG to GUG. In either case the change involved a single base in the several hundred needed to code for one of the two kinds of chain in hemoglobin.

The one exception so far to the rule that all amino acid changes could be caused by single base changes has been found by Yanofsky. In this one case glutamic acid was replaced by methionine. It can be seen from the genetic code dictionary that this can be accomplished only by a change of *two* bases, since glutamic acid is encoded by either GAA or GAG and methionine is encoded only by AUG. This mutation has occurred only once, however, and of all the mutations studied by Yanofsky it is the only one not to back-mutate, or revert to "wild type." It is thus almost certainly the rare case of a double change. All the other cases fit the hypothesis that base-substitution mutations are normally caused by a single base change. Examination of the code shows that only about 40 percent of all the possible amino acid interchanges can be brought about by single base substitutions, and it is only these changes that are found in experiments. Therefore the study of actual mutations has provided strong confirmation of many features of the genetic code.

Because in general several codons stand for one amino acid it is not possible, knowing the amino acid sequence, to write down the exact RNA base sequence that encoded it. This is unfortunate. If we know which amino acid is changed into another by mutation, however, we can often, given the code, work out what that base change must have been. As an example, glutamic acid can be encoded by GAA or GAG and valine by GUU, GUC, GUA or GUG. If a mutation substitutes valine for glutamic acid, one can assume that only a single base change was involved. The only such change that could lead to the desired result would be a change from A to U in the middle position, and this would be true whether GAA became GUA or GAG became GUG.

It is thus possible in many cases (not in all) to compare the nature of the base change with the chemical mutagen used to produce the change. If RNA is treated with nitrous acid, C is changed to U and A is effectively changed to G. On the other hand, if double-strand DNA is treated under the right conditions with hydroxylamine, the mutagen acts only on C. As a result some C's are changed to T's (the DNA equivalent of U's), and thus G's, which are normally paired with C's in double-strand DNA, are replaced by A's.

If 2-aminopurine, a "base analogue" mutagen, is added when double-strand DNA is undergoing replication, it produces only "transitions." These are the same changes as those produced by hydroxylamine—plus the reverse changes. In almost all these different cases (the exceptions are unimportant) the changes observed are those expected from our knowledge of the genetic code.

Note the remarkable fact that, although the code was deduced mainly from studies of the colon bacillus, it appears to apply equally to human beings and tobacco plants. This, together with more fragmentary evidence, suggests that the genetic code is either the same or very similar in most organisms.

The second method of checking the code using intact cells depends on phase-shift mutations such as the addition of a single base to the message. Phase-shift mutations probably result from errors produced during genetic recombination or when the DNA molecule is being duplicated. Such errors have the effect of putting out of phase the reading of the message from that point on. This hypothesis leads to the prediction that the phase can be corrected if at some subsequent point a nucleotide is deleted. The pair of alterations would be expected not only to change two amino acids but also to alter all those encoded by bases lying between the two affected sites. The reason is that the intervening bases would be read out of phase and therefore grouped into triplets different from those contained in the normal message.

This expectation has recently been confirmed by George Streisinger and his colleagues at the University of Oregon. They have studied mutations in the protein lysozyme that were produced by the T4 virus, which infects the colon bacillus. One phase-shift mutation involved the amino acid sequence ...Lys–Ser–Pro–Ser–Leu–AspN–Ala–Ala–Lys.... They were then able to construct by genetic methods a double phase-shift mutant in which the corresponding sequence was ...Lys–Val–His–His–Leu–Met–Ala–Ala–Lys....

Given these two sequences, the reader should be able, using the genetic code dictionary on page 96, to decipher uniquely a short length of the nucleic acid message for both the original protein and the double mutant and thus deduce the changes produced by each of the phase-shift mutations. The correct result is presented in the illustration above. The result not only confirms several rather doubtful codons, such as UUA for leucine and AGU for serine, but also shows which codons are actually involved in a genetic message. Since the technique is difficult, however, it may not find wide application.

Streisinger's work also demonstrates what has so far been only tacitly assumed: that the two languages, both of which are written down in a certain direction according to convention, are in fact translated by the cell in the same direction and not in opposite directions. This fact had previously been established, with more direct chemical methods, by Severo Ochoa and his colleagues at the New York University School of Medicine. In the convention, which was adopted by chance, proteins are written with the amino (NH_2) end on the left. Nucleic acids are written with the end of the molecule containing a "5 prime" carbon atom at the left. (The "5 prime" refers to a particular carbon atom in the 5-carbon ring of ribose sugar or deoxyribose sugar.)

Finding the Anticodons

Still another method of checking the genetic code is to discover the three bases making up the anticodon in some particular variety of transfer RNA. The first tRNA to have its entire sequence worked out was alanine tRNA, a job done by Robert W. Holley and his collaborators at Cornell University [see "The Nucleotide Sequence of a Nucleic Acid," by Robert W. Holley; SCIENTIFIC AMERICAN Offprint 1033]. Alanine tRNA, obtained from yeast, contains 77 bases. A possible anticodon found near the middle of the molecule has the sequence IGC, where I stands for inosine, a base closely resembling guanine. Since then Hans Zachau and his colleagues at the University of Cologne have established the sequences of two closely related serine tRNA's from yeast, and James Madison and his group

at the U.S. Plant, Soil and Nutrition Laboratory at Ithaca, N.Y., have worked out the sequence of a tyrosine tRNA, also from yeast.

A detailed comparison of these three sequences makes it almost certain that the anticodons are alanine–IGC, serine–IGA and tyrosine–GΨA. (Ψ stands for pseudo-uridylic acid, which can form the same base pairs as the base uracil.) In addition there is preliminary evidence from other workers that an anticodon for valine is IAC and an anticodon for phenylalanine is GAA.

All these results would fit the rule that the codon and anticodon pair in an antiparallel manner, and that the pairing in the first two positions of the codon is of the standard type, that is, A pairs with U and G pairs with C. The pairing in the third position of the codon is more complicated. There is now good experimental evidence from both Nirenberg and Khorana and their co-workers that one tRNA can recognize several codons, provided that they differ only in the last place in the codon. Thus Holley's alanine tRNA appears to recognize GCU, GCC and GCA. If it recognizes GCG, it does so only very weakly.

The "Wobble" Hypothesis

I have suggested that this is because of a "wobble" in the pairing in the third place and have shown that a reasonable theoretical model will explain many of the observed results. The suggested rules for the pairing in the third position of the anticodon are presented in the table at the top of this page, but

ANTICODON	CODON
U	A G
C	G
A	U
G	U C
I	U C A

"WOBBLE" HYPOTHESIS has been proposed by the author to provide rules for the pairing of codon and anticodon at the *third* position of the codon. There is evidence, for example, that the anticodon base I, which stands for inosine, may pair with as many as three different bases: U, C and A. Inosine closely resembles the base guanine (G) and so would ordinarily be expected to pair with cytosine (C). Structural diagrams for standard base pairings and wobble base pairings are illustrated at the bottom of this page.

this theory is still speculative. The rules for the first two places of the codon seem reasonably secure, however, and can be used as partial confirmation of the genetic code. The likely codon-anticodon pairings for valine, serine, tyrosine, alanine and phenylalanine satisfy the standard base pairings in the first two places and the wobble hypothesis in the third place [see *illustration on page 100*].

Several points about the genetic code remain to be cleared up. For example, the triplet UGA has still to be allocated.

GUANINE   CYTOSINE   GUANINE   URACIL

RIBOSE SUGAR   RIBOSE SUGAR   RIBOSE SUGAR   RIBOSE SUGAR

STANDARD AND WOBBLE BASE PAIRINGS both involve the formation of hydrogen bonds when certain bases are brought into close proximity. In the standard guanine-cytosine pairing (*left*) it is believed three hydrogen bonds are formed. The bases are shown as they exist in the RNA molecule, where they are attached to 5-carbon rings of ribose sugar. In the proposed wobble pairing (*right*) guanine is linked to uracil by only two hydrogen bonds. The base inosine (I) has a single hydrogen atom where guanine has an amino (NH_2) group (*broken circle*). In the author's wobble hypothesis inosine can pair with U as well as with C and A (*not shown*).

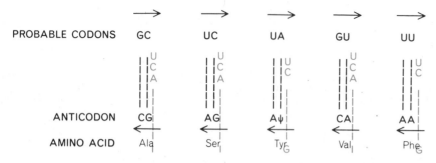

CODON-ANTICODON PAIRINGS take place in an antiparallel direction. Thus the anticodons are shown here written backward, as opposed to the way they appear in the text. The five anticodons are those tentatively identified in the transfer RNA's for alanine, serine, tyrosine, valine and phenylalanine. Color indicates where wobble pairings may occur.

The punctuation marks—the signals for "begin chain" and "end chain"—are only partly understood. It seems likely that both the triplet UAA (called "ochre") and UAG (called "amber") can terminate the polypeptide chain, but which triplet is normally found at the end of a gene is still uncertain.

The picturesque terms for these two triplets originated when it was discovered in studies of the colon bacillus some years ago that mutations in other genes (mutations that in fact cause errors in chain termination) could "suppress" the action of certain mutant codons, now identified as either UAA or UAG. The terms "ochre" and "amber" are simply invented designations and have no reference to color.

A mechanism for chain initiation was discovered fairly recently. In the colon bacillus it seems certain that formyl-methionine, carried by a special tRNA, can initiate chains, although it is not clear if all chains have to start in this way, or what the mechanism is in mammals and other species. The formyl group (CHO) is not normally found on finished proteins, suggesting that it is probably removed by a special enzyme. It seems likely that sometimes the methionine is removed as well.

It is unfortunately possible that a few codons may be ambiguous, that is, may code for more than one amino acid. This is certainly not true of most codons. The present evidence for a small amount of ambiguity is suggestive but not conclusive. It will make the code more difficult to establish correctly if ambiguity can occur.

Problems for the Future

From what has been said it is clear that, although the entire genetic code is not known with complete certainty, it is highly likely that most of it is correct. Further work will surely clear up the doubtful codons, clarify the punctuation marks, delimit ambiguity and extend the code to many other species. Although the code lists the codons that *may* be used, we still have to determine if alternative codons are used equally. Some preliminary work suggests they may not be. There is also still much to be discovered about the machinery of protein synthesis. How many types of tRNA are there? What is the structure of the ribosome? How does it work, and why is it in two parts? In addition there are many questions concerning the control of the rate of protein synthesis that we are still a long way from answering.

When such questions have been answered, the major unsolved problem will be the structure of the genetic code. Is the present code merely the result of a series of evolutionary accidents, so that the allocations of triplets to amino acids is to some extent arbitrary? Or are there profound structural reasons why phenylalanine has to be coded by UUU and UUC and by no other triplets? Such questions will be difficult to decide, since the genetic code originated at least three billion years ago, and it may be impossible to reconstruct the sequence of events that took place at such a remote period. The origin of the code is very close to the origin of life. Unless we are lucky it is likely that much of the evidence we should like to have has long since disappeared.

Nevertheless, the genetic code is a major milestone on the long road of molecular biology. In showing in detail how the four-letter language of nucleic acid controls the 20-letter language of protein it confirms the central theme of molecular biology that genetic information can be stored as a one-dimensional message on nucleic acid and be expressed as the one-dimensional amino acid sequence of a protein. Many problems remain, but this knowledge is now secure.

Symbiosis and Evolution

by Lynn Margulis
August 1971

The cells of higher plants and animals have specialized organelles such as chloroplasts and mitochondria. There is increasing reason to believe that these organelles were once independent organisms

Every form of life on earth—oak tree and elephant, bird and bacterium—shares a common ancestry with every other form; this fact has been conclusively demonstrated by more than a century of evolutionary research. At the same time every living thing belongs primarily to one or another of two groups that are mutually exclusive: organisms with cells that have nuclei and organisms with cells that do not. (An exception is viruses and virus-like particles, but such organisms can reproduce only inside cells.) How can both of these facts be true? Why does so profound a biological schism exist? Ideas put forward and discarded some decades ago hinted at one explanation: Cells without nuclei were the first to evolve. Cells with nuclei, however, are not merely mutant descendants of the older kind of cell. They are the product of a different evolutionary process: a symbiotic union of several cells without nuclei.

The cells of the two classes of organisms are called prokaryotic ("prenuclear") and eukaryotic ("truly nucleated"). The two classes are not equally familiar to us. Most of the forms of life we see—ourselves, trees, pets and the plants and animals that provide our food—are eukaryotes. Each of their cells has a central organelle: a membrane-enclosed nucleus where genetic material is organized into chromosomes. Each has within its cytoplasm several other kinds of organelle. Prokaryotes are far less prominent organisms, although they exist in huge numbers. In the absence of a membrane-enclosed nucleus their genetic material is dispersed throughout their cytoplasm. Such primitive simplicity is characteristic of the blue-green algae and of all the myriad species of bacteria.

The relatedness of living things is fundamental. Organisms as apparently dissimilar as men and molds have almost identical nucleic acids and have similarly identical enzyme systems for utilizing the energy stored in foodstuffs. Their proteins are made up of the same 20 amino acid units. In spite of a bewildering diversity of forms, in these fundamental respects living things are the same. Yet we are left with the equally fundamental discontinuity represented by the two different classes of cells.

Varieties of Symbiosis

Symbiosis can be defined as the living together of two or more organisms in close association. To exclude the many kinds of parasitic relationships known in nature, the term is often restricted to associations that are of mutual advantage to the partners. One frequently cited instance of symbiosis is the partnership sometimes observed between the hermit crab and the sea anemone. The anemone attaches itself to the shell that shelters the crab; this provides its partner with camouflage, and stray bits of the crab's food nourish the anemone. An example that is more pertinent here is the relationship between the leguminous plants and certain free-living soil bacteria. Neither organism can by itself utilize the gaseous nitrogen of the atmosphere. The roots of the plants, however, develop projections known as infection threads that transport the soil bacteria into the root structure. Once present in the cytoplasm of the root cells, the bacteria (transformed into "bacteroids") combine with the host cells to form a specialized tissue: the root nodule. Inert atmospheric nitrogen is utilized by nodule cells as a nutrient. At the same time the nodules manufacture a substance—a pinkish protein known as leghemoglobin—that neither the plant nor the bacteria alone can produce. Because

the bacterial symbionts live within the tissue of the plant host the partnership is classified as "endosymbiosis."

Neither of these relationships is necessarily hereditary. The hermit crab will never give rise to the anemone, nor the anemone to the hermit crab. Nor in most instances does a pea or an alfalfa seed contain bacteria; each new generation of plants must establish its own association with a new generation of bacteria. On the other hand, there is one plant—*Psychotria bacteriophila*—that contains the bacterial symbiont in its seed. Thus its offspring inherit not only chromosomes and cytoplasm from the parent plants but bacteria as well. This constitutes hereditary endosymbiosis.

Hereditary symbiosis is surprisingly common. In many instances the host—plant or animal—cannot manufacture its own food and the guest belongs to the family of organisms that can synthesize nutrients by absorbing sunlight. Hosts of this kind are heterotrophs: "other-feeders." Among plants the fungi fit into this group; so do most forms of animal life. Their guests are autotrophs: "self-feeders." The process that nourishes them is the familiar one we call photosynthesis.

An instance of such a relationship is provided by lichens, the characteristically flat, crusty plants that can survive in harshly dry and cold environments. Microscopic study long ago demonstrated that a lichen is a symbiotic partnership between an alga (the autotroph) and a fungus (the heterotroph). Vernon Ahmadjian of Clark University has managed to dissociate the partners that form lichens of the genus *Cladonia*, and he has succeeded in raising the two components independently.

Endosymbiosis has been characterized as swallowing without digesting. One protozoan symbiont—*Paramecium bursaria*, commonly known as the green paramecium—provides an apt illustration. This protozoon has been studied intensively by Richard Siegel of the University of California at Los Angeles and Stephen Karakashian of the State University of New York at Old Westbury. It is green because numerous photosynthetic green algae inhabit its single cell. The photosynthetic guests, given adequate light, can keep the host alive under near-starvation conditions. When the host is deprived of its guests, it will survive only if extra nutrients are added to its medium. The guests (members of the genus *Chlorella*, a common green alga) will also survive when they are removed from the host.

When the organism is reconstituted in the laboratory by bringing the isolated paramecium and the algae together, an interesting thing happens. Once back inside the host, the algae multiply, but only until the normal, genetically regulated number of algae per paramecium is attained. The multiplication then stops. Should the protozoon encounter free-living *Chlorella*, they are promptly digested. Its own algal partners, however, are totally immune. Somehow the paramecium recognizes its symbiont, although even with the electron microscope it is not easy to see any morphological difference between the free-living *Chlorella* and the symbiotic one.

The relationships described thus far involve hosts whose guests all belong to a single species. Far more complex kinds of symbiosis are known. There is one protozoon, for example, that is itself a symbiont and at the same time is the host of three other symbionts. This is the flagellate *Myxotricha paradoxa*, a large, smooth-swimming single-celled organism that seems to be covered with hair-like flagella of various sizes. *Myxotricha* lives in the gut of certain Australian termites; it contributes to the insects' survival by helping them digest the pulverized wood that comprises their food.

When *Myxotricha* was first described, it was thought to be just another multi-flagellate protozoon with an unusual mode of swimming.

A detailed study by A. V. Grimstone of the University of Cambridge and L. R. Cleveland of the University of Georgia revealed that *Myxotricha* actually had only a few normal flagella at one end. What were mistaken for flagella elsewhere on the organism were spirochetes—a kind of elongated motile bacterium—that were living symbiotically on the surface of the protozoan host. This was not all; each spirochete was associated with another kind of symbiotic bacterium that was also attached to the host's surface, and still a third kind of symbiotic bacterium lived inside *Myxotricha* [see illustration on page 104]. As Grimstone and Cleveland have noted, the protozoon "glides along uninterruptedly" through the gut of the termite "at constant speed and usually in a straight line," with its symbiotic spirochetes undulating vigorously.

Organelles of the Eukaryotic Cell

Having seen how many different kinds of independent organism can enter into symbiotic partnerships and how some of these partnerships can be perpetuated on a hereditary basis, we now turn to the eukaryotic cell. When we examine such a cell under the microscope, we see that it contains not only a nucleus but also other organelles. In the eukaryotic cells of a green leaf, for example, there are tiny green chloroplasts, where the chemical events of photosynthesis take place. In the cells of both plants and animals there are mitochondria, where foodstuffs are oxidized to produce ATP (adenosine triphosphate), the universal fuel of biochemical reactions. These are only two of several types of organelle.

Could these organelles have originated as independent organisms? One kind

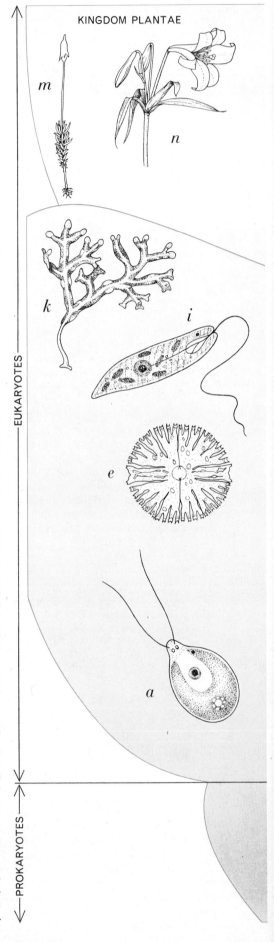

"FIVE-KINGDOM" CLASSIFICATION of terrestrial life, proposed by R. H. Whittaker of Cornell University to solve the dilemma posed by the conventional classification of organisms as either plants or animals, is shown as modified by the author. The life forms comprise two unambiguous and mutually exclusive groups: prokaryotes, the organisms with cells that lack membrane-enclosed nuclei, all within the kingdom Monera, and the eukaryotes, the organisms with truly nucleated cells, which include the populations of the other four. Organisms representative of major phyla are illustrated. In the kingdom Monera these are various bacteria (*left*) and a blue-green alga, *Nostoc* (*right*). In the kingdom Protista are *Chlamydomonas*, one of the chlorophyta (*a*), diatoms (*b*), an amoeba (*c*), a dinoflagellate (*d*), a desmid (*e*), a foraminiferan (*f*), a trypanosome (*g*), a sun animalcule (*h*), a euglena (*i*), a paramecium (*j*), a brown seaweed (*k*) and a cellular slime mold (*l*). The two phyla in the kingdom Plantae, a group nourished by photosynthesis, are represented by a haircap moss (*m*) and a lily (*n*). In the kingdom Fungi, a group characterized by absorptive nutrition, are a bread mold (*o*) and a mushroom (*p*). In the kingdom Animalia, characterized by ingestive nutrition, the representatives are a mollusk (*q*), arthropod (*r*) and chordate (*s*).

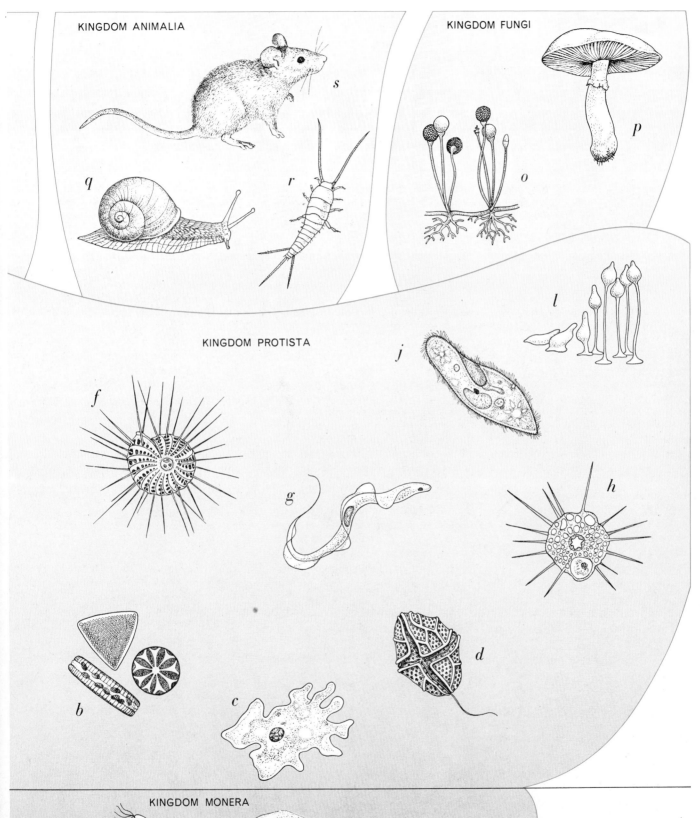

KINGDOM ANIMALIA

s

q

r

KINGDOM FUNGI

o

p

KINGDOM PROTISTA

j

l

f

g

h

b

c

d

KINGDOM MONERA

of evidence immediately suggests such an origin: the existence of what are known as cytoplasmic genes. When we speak of genes, we usually have in mind the hereditary material—the DNA—in the chromosomes of the cell nucleus. Yet genes are also found outside the nucleus in the cytoplasm, notably in association with chloroplasts and mitochondria.

Chloroplasts belong to a group of organelles collectively known as plastids. Plastids have their own unique DNA—a DNA unrelated to the DNA of the cell nucleus. As has been abundantly demonstrated over the past two decades, DNA is the replicative molecule of the cell. It encodes the synthesis of the proteins required for the doubling of the cell material before cell division. It has also been demonstrated that chloroplasts have their own ribosomes: the bodies where protein is synthesized. The present picture of cellular protein synthesis is that the hereditary information encoded in DNA is transcribed in "messenger" RNA, which then provides the ribosome with the information it needs to link amino acids into a particular protein. In the process each amino acid molecule temporarily combines with a specific molecule of another kind of RNA: "transfer" RNA. Chloroplasts also contain specific transfer RNA's and other components necessary for independent protein synthesis.

Mitochondria also contain DNA that is not related to the DNA of the cell nucleus. The mitochondria in animal cells apparently have only enough DNA and the associated protein-synthesizing machinery to produce a fraction of the structural protein and enzymes needed by these organelles in order to function. Nonetheless, the machinery is there: DNA, messenger RNA, special mitochondrial ribosomes and so forth. The presence of DNA associated with protein synthesis implies that the mitochondria have a functional genetic system.

Here, then, are two organelles of eukaryotic cells that have their own genes and conduct protein synthesis. When one considers that almost all the protein synthesis in the eukaryotic cell is under the direction of nuclear DNA and that the synthesis is accomplished by ribosomes in the cytoplasm external to both the mitochondria and the plastids, it is natural to wonder why these organelles carry duplicate equipment. Does their ability to grow and divide within the cell and to make some of their own protein under the direction of their own genes imply that they were once free-living organisms? A number of investigators have thought so.

When the plastids of eukaryotic algae were studied under the microscope in the 19th century, it was remarked that they resembled certain free-living algae, and it was suggested that they had originated as such algae. A similar origin for mitochondria was proposed in the 1920's by an American physician, J. E. Wallin. On the basis of microscopic observations, of reactions to stains and of assertions (subsequently refuted) that he had grown isolated mitochondria in the laboratory, Wallin maintained that mitochondria were bacteria that had come to live symbiotically within animal cells. In his book *Symbioticism and the Origin of Species* he argued that new species arise as a result of this kind of symbiosis between distantly related organisms. As can happen to people obsessed by a novel concept, Wallin overstated his case and used doubtful data to defend it. His book fell into disrepute.

What is known today about the biochemical autonomy of mitochondria goes a long way toward rehabilitating Wallin's basic concept. It now seems certain that mitochondria were once free-living bacteria that over a long period of time established a hereditary symbiosis with ancestral hosts that ultimately evolved into animal cells, plant cells and cells that fit neither of these categories. The same history evidently holds true for plastids, which were originally free-living algae. I believe that still a third group of organelles, the flagella and cilia, became associated with the eukaryotic cell in much the same way.

Flagella and Cilia

Flagella and cilia are really the same. If these hairlike cell projections are long and few, they are called flagella; if they are short and many, they are called cilia. Their motion propels the cell through its medium or, if the cell is fixed in place, moves things past it. In the tissues of higher animals some flagella and cilia have been drastically modified to serve other functions. The light receptors in the eye of vertebrates are such structures. So are the smell receptors of vertebrates. Among prokaryotes the analogous structures are much simpler. They are small, single-stranded and consist of a protein called flagellin.

The flagella and cilia of eukaryotic cells are much larger than those of prokaryotes. Their basic structure is strikingly uniform, whether they come from the sperm of a fern or the nostril of a mouse. Seen in cross section, each consists of a circle of paired microtubules surrounding one centrally located pair. If the structure is motile, there are always two microtubules in the middle and always nine more pairs surrounding them; the pattern is known as the "9 + 2 array" [see illustrations on page 106]. Microtubules from any kind of eukaryotic flagella and cilia are composed of related proteins called tubulin.

At the base of every eukaryotic flagellum and cilium is a distinct microtubular structure: the basal body. The architecture of the basal body is identical with that of the centriole, a structure found

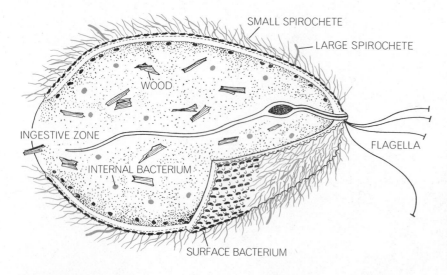

SMALL SPIROCHETE
LARGE SPIROCHETE
WOOD
INGESTIVE ZONE
INTERNAL BACTERIUM
FLAGELLA
SURFACE BACTERIUM

COMPLEX SYMBIONT, the protozoon *Myxotricha paradoxa*, lives as a guest in the gut of certain Australian termites and plays host to three symbionts of its own. These are surface bacteria of the spirochete group (*color*), which observers first mistook for flagella, other surface bacteria (*black*) and still other bacteria (*color*) that live inside the protozoon.

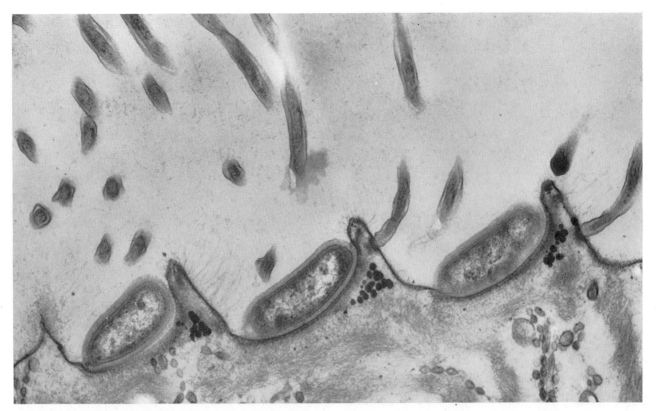

SURFACE OF *MYXOTRICHA* appears at the bottom in transverse section in this electron micrograph by A. V. Grimstone of the University of Cambridge. In the "hollow" to the left of each surface "peak" lies one of the bacterium guests of the protozoan host. Two symbiotic spirochetes are visible at right with their basal ends attached to the host membrane. Other spirochetes whose attachments are not in the plane of focus are partially visible elsewhere in the micrograph (*top*). The theory proposing that eukaryotic cells are the products of similar symbiotic relationships suggests that the first symbionts were free-living bacterium-like prokaryotes.

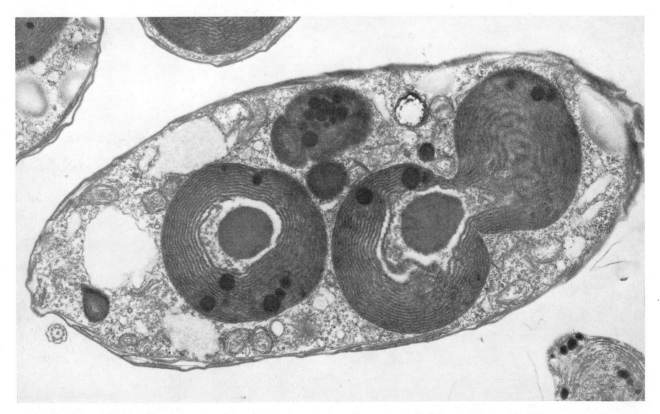

PROKARYOTIC GUESTS, identifiable by their array of concentric photosynthetic membranes as the blue-green alga *Cyanocyta*, are enlarged 15,000 times in this electron micrograph by William T. Hall of the National Institutes of Health. They are inside protozoan hosts of the species *Cyanophora paradoxa*. Similar hereditary symbioses between various photosynthetic alga-like prokaryotes and large, more advanced eukaryotic hosts from the kingdom Protista is suggested as the step leading to evolution of the plant kingdom.

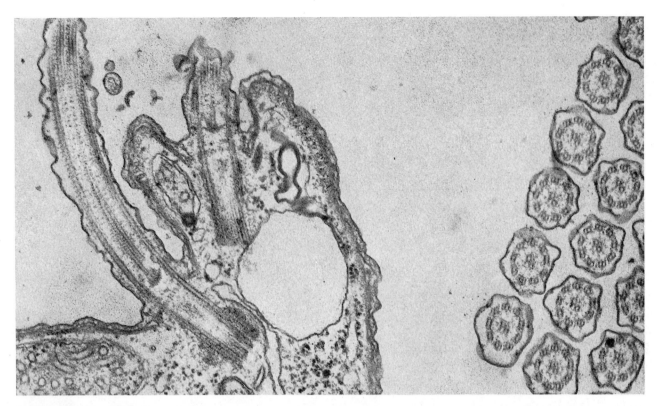

STRUCTURE OF FLAGELLA is shown in transverse section (*right*) and longitudinal section (*left*) in an electron micrograph by R. D. Allen of the University of Hawaii. In the part of the flagellum extending beyond the basal body a circle of paired microtubules surrounds a central pair in what is known as a "9 + 2 array." In the basal body the central pair of microtubules is absent and the array is "9 + 0." Such organelles are found only among the eukaryotes and may originally have been free-living cells.

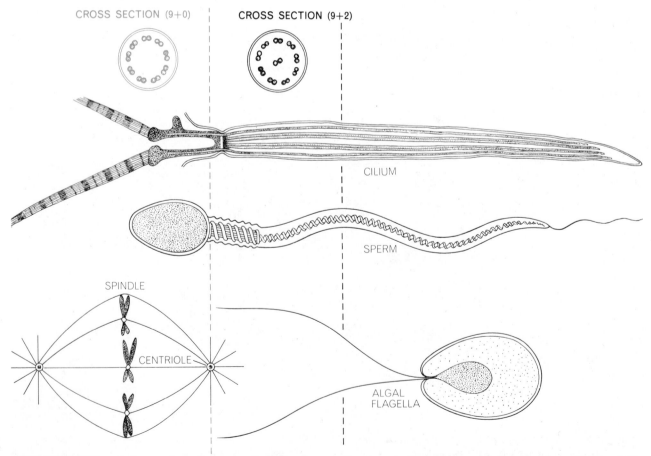

MICROTUBULES comprise a variety of structures, including the motile flagella of certain algae (*bottom right*) and of sperm, the cilia of tracheal membrane and the centrioles and the spindle structure that mediates halving of the nucleus in mitotic division.

at opposite poles of the eukaryotic cell nucleus. Centrioles come into particular prominence during mitosis, the process by which eukaryotic cells divide. (Centrioles are found in nearly all animal cells and in the cells of many eukaryotic algae but not in certain fungi and in most higher plants).

The structural array of the basal body and the centriole is "9 + 0": the central pair of microtubules is absent. In cells that possess mitotic centrioles the centrioles left over from earlier cell divisions often grow projections that become flagella or cilia as the new cell differentiates. Thus not only are basal bodies and centrioles identical in structural pattern but also centrioles can become basal bodies. Moreover, the mitotic spindle, the characteristic diamond-shaped structure that lies between the centrioles during cell division, is an array of microtubules composed of tubulin.

A further finding requires that we now ask two fundamental questions. When the plant alkaloid colchicine is added to tubulin, derived either from flagella or cilia or from spindles, the alkaloid is bound to the protein. The reaction is characteristic of tubulin from the cells of all animals and all eukaryotic plants, but it has never been observed with the flagellin from prokaryotic cells. Nor, for that matter, have microtubules ever been observed in either bacteria or blue-green algae.

The first question is this: What differentiates animals from plants? At the macroscopic level the differences are obvious; for example, most animals move around in order to feed themselves, whereas most plants stand still and nourish themselves by photosynthesis. At the microscopic level distinctions of this kind become meaningless. Many kinds of single-celled organism sometimes nourish themselves by photosynthesis and at other times swim about ingesting food particles. Some organisms crawl like an amoeba at one stage in their development but later stop, sprout stems and disperse a new generation in the form of spores. Further examples are almost innumerable.

Generations of biologists have been troubled by the need to force such organisms into the plant or animal kingdom. A far less ambiguous dichotomy is the division between prokaryotes and eukaryotes. Notable dissenters from the plant-animal classification are Herbert F. Copeland of Sacramento State College, G. Evelyn Hutchinson of Yale University and most recently R. H. Whittaker of Cornell University. In what follows I have modified Whittaker's "five-

kingdom" classification, which takes the fundamental prokaryote-eukaryote dichotomy fully into account [see illustration on pages 102 and 103]. The answer to the first question, then, is that there are not just two basic kinds of organism but five.

This brings us to the second question: How did five kingdoms arise? I have already suggested that eukaryotic cells, which are characteristic of all higher forms of life, came into existence through an evolutionary advance of a kind fundamentally different from discrete mutation. Specific answers to the second question will appear in the following hypothetical reconstruction of the origin of eukaryotic cells. The reconstruction traces the rise of the more advanced four of Whittaker's five kingdoms from their origin in the least advanced one. That kingdom is the kingdom Monera: the prokaryotic single-celled organisms that were the first living things to evolve on the earth. The reader should be warned that my presentation of the theory here is necessarily brief and oversimplified.

The First Cells

All life on the earth is believed to have originated more than three billion years ago during Lower Precambrian times in the form of bacterium-like prokaryotic cells. At that time there was no free oxygen in the atmosphere. The cells that arose were fermenting cells; their food consisted of organic matter that had been produced earlier by the action of various abiotic processes. Under pressures of natural selection directly related to the depletion of this stock of abiotic nutrients, there arose among the first fermenting bacteria many metabolic traits that are still observable among bacteria living today. These traits include the ability to ferment many different carbohydrates, to incorporate atmospheric carbon dioxide directly into reduced metabolic compounds, to reduce sulfate to hydrogen sulfide as a by-product of fermentation, and so on.

As the ammonia available in some parts of the environment became depleted, certain bacteria evolved metabolic pathways that could "fix" atmospheric nitrogen into amino acids. Other fermenters developed into highly motile organisms that foreshadowed such highly mobile living bacteria as spirochetes. All these fermenting bacteria were "obligate anaerobes," that is, for them oxygen was a powerful poison. Through various detoxification mechanisms the fermenters were able to dispose of the small amount

of this deadly element present in the environment as a result of abiotic processes. Finally, many if not all of the various fermenting bacteria were equipped with well-developed systems for the repair of DNA. Such systems were necessary to counteract the damage done by ultraviolet radiation, which at that time was intense because there was no ozone (O_3) in the atmosphere to filter it out.

All these bacteria were heterotrophs; they had not evolved the photosynthetic mechanisms that would have enabled them to nourish themselves in the absence of abiotic organic compounds. In time some of them developed metabolic pathways that led to the synthesis of the compounds known as porphyrins. It is a purely fortuitous property of porphyrins that they absorb radiation at the visible wavelengths; nonetheless, this property was eventually put to use by the evolution of bacterial photosynthesis. The process of photosynthesis requires a source of hydrogen. Bacteria can utilize such inorganic substances as hydrogen sulfide and gaseous hydrogen as well as various organic compounds of the kind that would have been present in the environment as by-products of fermentation. These first anaerobic photosynthesizers appeared in Lower Precambrian times.

When the new photosynthetic bacteria became well established, a process that may have taken millions of years, a second kind of photosynthesis was able to make its appearance. In the second process the uptake of hydrogen was accomplished by the splitting of water molecules; as a result increasing quantities of lethal free oxygen entered the atmosphere as a waste product. The evolution of this mode of photosynthesis led to the appearance of the blue-green algae, the first organisms on the earth that were adapted to the presence of free oxygen. Since they were active photosynthesizers of the newer type, they accelerated the increase in atmospheric oxygen.

The blue-green algae, whose Precambrian success is attested by the massive calcium-rich rock formations they left behind, presented a profound threat to all other forms of life. The other organisms were forced to adapt or perish. Some of the anaerobes adapted simply by retreating into the oxygen-free muds where their fellows are found today. Others developed new mechanisms of oxygen detoxification; still others, it is safe to assume, merely disappeared. In any case, one result of the success of the blue-green algae was the evolution of new kinds of bacteria that utilized free oxygen in their metabolic processes: aerobic respirers, oxidizers of sulfide and ammonia, and the like. As atmospheric oxygen continued to accumulate, the stage was set for the initial appearance of eukaryotic cells.

The First Eukaryote

The first advanced cell came into existence when some kind of host, perhaps a fermenting bacterium, acquired as symbiotic partners a number of smaller oxygen-respiring bacteria. As atmospheric oxygen continued to increase, selection pressure would have favored such a symbiosis. Eventually the small aerobic bacteria became the hereditary guests of their hosts; these were the first mitochondria. The host symbionts, in turn, evolved in the direction of amoebas, so that a new population of large aerobic cells evolved and faced the problem of finding nutrients.

In due course the partners were aided in their quest for food: a second group of symbionts, flagellum-like bacteria comparable to modern spirochetes, attached themselves to the host's surface and greatly increased its motility. If this hypothetical triple partnership begins to resemble the termite symbiont *Myxotricha*, it is with good reason; I believe that just such a *Myxotricha*-like symbiotic association, formed in Precambrian times, was a universal ancestor to all eukaryotic organisms. With the appearance of this supercell the kingdom Monera gives rise, in a manner consistent with Whittaker's taxonomic system, to the kingdom Protista.

The internal guests, then, served as mitochondria and the external ones as flagella. The spirochete-like guests, however, slowly evolved another role. The specialized basal body of the flagellum and its associated microtubules came to serve the additional function of mediating the process of cell division. Respectively the centriole and the mitotic spindle, they were responsible for dividing the parent cell's genes evenly between daughter cells.

Mitotic cell division was the crucial genetic step toward further evolutionary advance. One would not expect it to have developed in a straight-line manner, starting with no mitosis and concluding with perfect mitosis. There must have been numerous dead ends, variations and byways. Evidence of just such uncertain gradualism is found today among the lower eukaryotes, for example the slime molds, the yellow-green and golden-yellow algae, the euglenids, the slime-net amoebas and others. Many of their mitotic arrangements are unconventional. The perfection of mitosis may have occupied as much as a billion years of Precambrian time.

Mitosis, however, was the key to the future. Without mitosis there could be no meiosis, the type of cell division that gives rise to eggs and sperm. There could be no complex multicellular organisms and no natural selection along Mendelian genetic lines. As mitosis was perfected the kingdom Protista gave rise to three other new kingdoms.

Plant-like protists probably appeared several times through symbiotic unions between free-living, autotrophic prokaryote blue-green algae and various heterotrophic eukaryote protists. After much modification the guest algae developed into those key organelles of the plant kingdom, the photosynthetic plastids. Some of the original symbiotic organisms are represented today by the eukaryotic algae that eventually evolved into the ancestors of the plant kingdom. Both algae with nucleated cells and higher plants have of course evolved a great deal since they first acquired photosynthetic guest plastids more than half a billion years ago. Their evolutionary progress, however, involves neither the origin nor any fundamental modification of the photosynthetic process. This heritage from their anaerobic prokaryote ancestors they received fully formed at the close of the Precambrian.

The group of organisms that we know as the fungi—molds, mushrooms, yeasts and the like—are also thought to derive directly from protists that relinquished flagellar motility in exchange for mitosis. This suggestion is consistent with Whittaker's classification. He splits the fungi from the plant kingdom and recognizes that these fundamentally different organisms deserve a domain of their own. The evolution of the animal kingdom, in turn, is considered a straight-line consequence of natural selection acting on the multicellular, sexually reproductive organisms that, like the fungi, did not happen to play host to plastids in Upper Precambrian times.

Testing the Hypothesis

Compared with what had gone before, however, all this seems to be virtually modern history. It is more pertinent at this juncture to see if the theory of

SYMBIOSIS THEORY is summarized in the three steps illustrated here. Union between two members of the kingdom Monera, a newly evolved aerobic bacterium (*bottom left*) and a larger host, possibly a fermenting bacterium (*bottom right*), brought into existence an amoeboid-like protist whose several guests became mitochondria. A second hereditary symbiosis, joining the amoeboid to a bacterium of the spirochete group (*center right*), brought into being an ancestral "amoeboflagellate" that was the direct forebear of two kingdoms: Fungi and Animalia. When the same amoeboflagellate went on to form another relationship, with algae that became plastids, the fifth kingdom, Plantae, was founded.

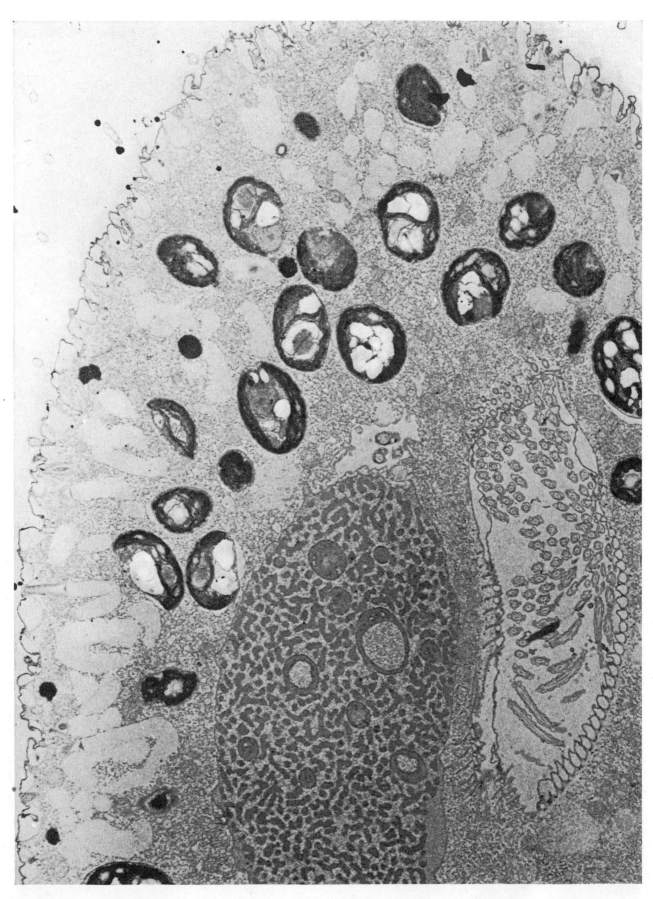

HEREDITARY SYMBIOSIS between photosynthetic green algae of the genus *Chlorella* (*scattered dark ovals*) and a single-celled animal host is characteristic of the species *Paramecium bursaria*, seen magnified 8,000 times in this electron micrograph. Even when the host is kept close to starvation, its guest symbionts satisfy its basic food requirements as long as sunlight is available. The chloroplasts in photosynthetic cells may once have been similar free-living alga-like organisms that eventually became guest symbionts.

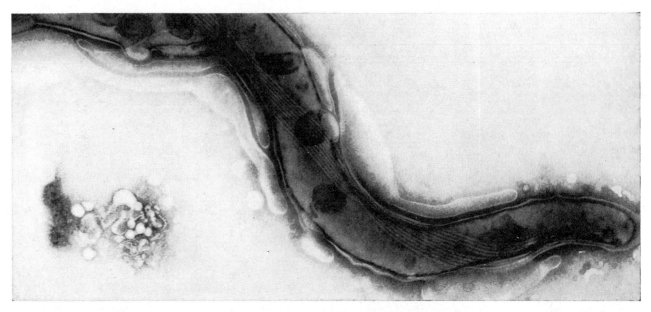

FREE-LIVING MONERAN, a bacterium of the spirochete group, is seen magnified 55,000 times in this electron micrograph. It is an anaerobic bacterium found in the human mouth. Organisms like this may have given rise to the eukaryotic flagella through symbiosis.

eukaryotic-cell origin through hereditary symbiosis offers useful answers to further outstanding questions.

Why are there genes outside cell nuclei? Some cytoplasmic genes may have arisen in other ways, but the symbiosis theory holds that the genes associated with chloroplasts and mitochondria demonstrate that these two kinds of organelle were once free-living organisms.

Why does evidence for photosynthesis appear in Middle Precambrian times, even though no higher plants appear in the fossil record until a mere 600 million years ago? The theory proposes that the higher plants are the result of a symbiosis between animal-like hosts and photosynthetic blue-green-alga-like guests whose partnership could not have evolved until relatively recent times, when mitosis had been perfected.

Why should there be any connection

between, on the one hand, the basal body and the flagellum and, on the other, the centriole and the mitotic spindle? The proposal is that the original free-living organism that once accounted only for the function of motility was ancestral to the organelles that came to mediate the equal partition of genetic material between daughter cells during mitosis.

Obviously many other questions remain to be answered. Can the synthesis of DNA and of messenger RNA be detected in association with the reproduction of the basal body and the centriole? Can evidence be found of a unique protein-synthesis system associated with these bodies? Without such evidence the case for these organelles having once been free-living organisms is weak. How and when did meiosis evolve from mitosis? Which organisms were the initial

hosts to the guest bacteria that became mitochondria? Were guest plastids of different kinds—red, brown, golden-yellow—acquired independently by the various kinds of eukaryotic algae? One related question is profoundly social. Can botanists, invertebrate zoologists and microbiologists, with their widely different backgrounds, agree on a single classification and a consistent evolutionary scheme for the lower organisms?

Conclusive proof that the symbiosis theory is correct demands experiment. The symbiotic partners will have to be separated, grown independently and then brought back into the same partnership. No organelle of a eukaryotic cell has yet been cultivated outside the cell. The function of a theory, however, is to make reasonable predictions that can be proved or disproved. The predictions of the symbiosis theory are clear.

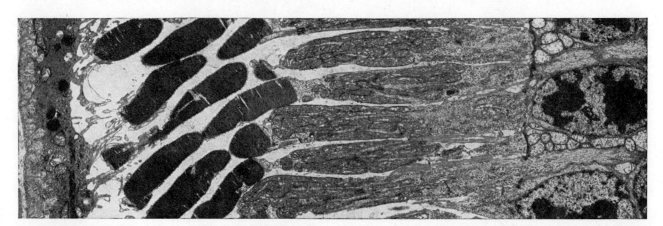

SPECIALIZED FLAGELLA, shown enlarged 3,900 times in an electron micrograph by Toichiro Kuwabara of the Harvard Medi- cal School, are visual receptors in the retina of a rabbit. The darker structures at left are the outer segments of the visual receptors.

11

Computer Analysis of Protein Evolution

by Margaret Oakley Dayhoff
July 1969

Amino acid sequences of similar proteins in different organisms contain information on relations among species. This information is analyzed to reconstruct in detail the history of living things

The protein molecules that determine the form and function of every living thing are intricately folded chains of amino acid units. The primary structure of each protein—the sequence in which its amino acid units are linked together—is governed by the sequence of subunits in the nucleic acid of the genetic material. The proteins of an organism are therefore the immediate manifestation of its genetic endowment. From a biochemical point of view a fungus and a man are different primarily because each of them has a different complement of proteins.

Yet human beings and fungi and organisms of intermediate biological complexity have some proteins in common. These homologous proteins are quite similar in structure, reflecting the ultimate common ancestry of all living things and the remarkable extent to which proteins have been conserved throughout geologic time. Because of this conservation the millions of proteins existing today are in effect living fossils: they contain information about their own origin and history and about the ancestry and evolution of the organisms in which they are found. The comparative study of proteins therefore provides an approach to critical issues in biology: the exact relation and order of origin of the major groups of organisms, the evolution of the genetic and metabolic complexity of present-day organisms and the nature of biochemical processes. A new discipline, chemical paleogenetics, concerns itself with such studies [see "The Evolution of Hemoglobin," by Emile Zuckerkandl; SCIENTIFIC AMERICAN, Offprint 1012]. In order to exploit the possibilities of this new field we have developed a computer technique for analyzing the relations among protein sequences.

The body of data available in protein sequences is something fundamentally new in biology and biochemistry, unprecedented in quantity, in concentrated information content and in conceptual simplicity. The data give direct information about the chemical linkage of atoms, and that linkage determines how protein chains coil, fold and cross-link—and thus establishes the three-dimensional structure of proteins. Because of our interest in the theoretical aspects of protein structure our group at the National Biomedical Research Foundation has long maintained a collection of known sequences. For the past four years we have published an annual *Atlas of Protein Sequence and Structure*, the latest volume of which contains nearly 500 sequences or partial sequences established by several hundred workers in various laboratories. In addition to the sequences, we include in the *Atlas* theoretical inferences and the results of computer-aided analyses that illuminate such inferences. This article is based in part on that material, to which contributions have been made by Chan Mo Park, Minnie R. Sochard, Lois T. Hunt and Patricia J. McLaughlin, and by Richard V. Eck, now of the University of Georgia.

Basic metabolic processes are similar in all living cells. Many identical structures, mechanisms, compounds and chemical pathways are found in widely diverse organisms; even the genetic code is almost the same in all species. It is by this code that the sequence of nucleotides, or nucleic acid subunits, that constitutes a gene is translated into the amino acid sequence of the protein derived from it. It is therefore not surprising that a large number of proteins have been found to have identifiable counterparts in most living things. These homologues appear to perform the same functions in the organisms in which they are found, and they can often be substituted for one another in laboratory experiments. Being complex substances, they are only rarely identical, but in the past 15 years homologous proteins have been shown to have nearly the same amino acid sequences and quite similar three-dimensional structures.

One such protein whose amino acid sequence has been established for a number of species is the protein of cytochrome c, a complex substance that in animals and higher plants is found in the cellular organelles called mitochondria, where it plays a role in biological oxidation. Twenty different sequences of cytochrome c have been identified and analyzed by a number of investigators, including Emil L. Smith of the University of California at Los Angeles, Emanuel Margoliash of the Abbott Laboratories and Shung Kai Chan and I. Tulloss of the University of Kentucky. Recently Karl M. Dus, Knut Sletten and Martin D. Kamen of the University of California at San Diego found a clearly related protein in a bacterium, *Rhodospirillum rubrum*, which lacks mitochondria.

The correspondence in amino acid sequence among these proteins is clear when the sequences are arrayed below one another [*see top illustration on next two pages*]. There are differences in length, reflecting additions or deletions of nucleotides in the corresponding genes. These changes are at the ends of sequences except for the internal deletions or additions revealed by the bacterial protein. Once the sequences have been adjusted to allow for these changes there is no question about the correct alignment. In man and the gray kangaroo, for example, the amino acids are the same in 94 out of 104 positions; in the less similar human and baker's yeast sequences, 64 positions conform, or some

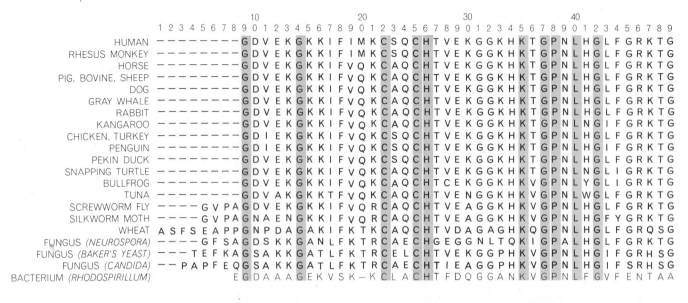

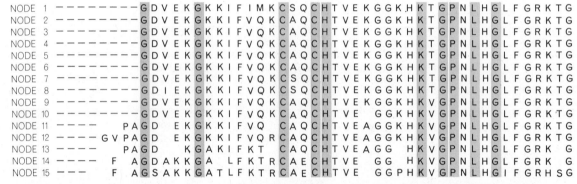

A ALANINE
C CYSTEINE
D ASPARTIC ACID
E GLUTAMIC ACID
F PHENYLALANINE
G GLYCINE
H HISTIDINE

I ISOLEUCINE
K LYSINE
L LEUCINE
M METHIONINE
N ASPARAGINE
P PROLINE
Q GLUTAMINE

R ARGININE
S SERINE
T THREONINE
V VALINE
W TRYPTOPHAN
Y TYROSINE

AMINO ACID SEQUENCES of 20 cytochrome c proteins and of a related bacterial protein are arrayed below one another. For the purposes of the computer each amino acid is represented by a single letter (see key at left) instead of the usual three-letter symbol. The proteins differ in length, and dashes have been inserted in order to preserve the correct alignment; these differences come at

three-fifths of the total length. All 21 sequences, including the bacterial one, have the same amino acid in 20 positions. When the amino acids at a given position are not the same, they usually have similar shapes or chemical properties.

Such similarity of sequence is impressive testimony to the evolution of all these organisms from common ancestors, confirming earlier morphological, embryological and fossil evidence. The alternative to common ancestry—that the similar cytochrome c proteins originated independently in different organisms—is not plausible. Consider the probability

of duplicating the sequence of amino acids in just one chain 100 units long. Since any of 20 amino acids can occur in every position, the number of different possible chains is 20^{100}. With so many possibilities it is improbable that two unrelated organisms would happen independently to have manufactured—

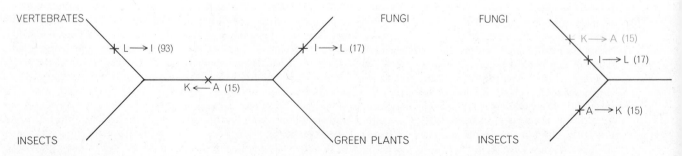

LINEAGES of major groups are constructed from the evidence at three positions in the sequences in order to illustrate the principles involved in constructing a phylogenetic tree. At Position 15 vertebrates and insects have a K (lysine); fungi and wheat have an A (alanine). This suggests that a single lineage connected the animals and plants and that a single mutation from A to K took

```
0 1 2 3 4 5 6 7 8 9 0 1 2 3 4 5 6 7 8 9 0 1 2 3 4 5 6 7 8 9 0 1 2 3 4 5 6 7 8 9 0 1 2 3 4 5 6 7 8 9 0 1 2 3 4 5 6 7 8 9 0 1 2

Q A P G Y S Y T A A N K N K G I I W G E D T L M E Y L E N P K K Y I P G T K M I F V G I K K K E E R A D L I A Y L K K A T N E
Q A P G Y S Y T A A N K N K G I T W G E D T L M E Y L E N P K K Y I P G T K M I F V G I K K K E E R A D L I A Y L K K A T N E
Q A P G F T Y T D A N K N K G I T W K E E T L M E Y L E N P K K Y I P G T K M I F A G I K K K T E R E D L I A Y L K K A T N E
Q A P G F S Y T D A N K N K G I T W G E E T L M E Y L E N P K K Y I P G T K M I F A G I K K K G E R E D L I A Y L K K A T N E
Q A P G F S Y T D A N K N K G I T W G E E T L M E Y L E N P K K Y I P G T K M I F A G I K K T G E R A D L I A Y L K K A T K E
Q A V G F S Y T D A N K N K G I T W G E E T L M E Y L E N P K K Y I P G T K M I F A G I K K K G E R A D L I A Y L K K A T N E
Q A V G F S Y T D A N K N K G I T W G E D T L M E Y L E N P K K Y I P G T K M I F A G I K K K D E R A D L I A Y L K K A T N E
Q A P G F T Y T D A N K N K G I I W G E D T L M E Y L E N P K K Y I P G T K M I F A G I K K K G E R A D L I A Y L K K A T N E
Q A E G F S Y T D A N K N K G I T W G E D T L M E Y L E N P K K Y I P G T K M I F A G I K K K S E R V D L I A Y L K D A T S K
Q A E G F S Y T D A N K N K G I T W G E D T L M E Y L E N P K K Y I P G T K M I F A G I K K K S E R A D L I A Y L K D A T S K
Q A E G F S Y T D A N K N K G I T W G E D T L M E Y L E N P K K Y I P G T K M I F A G I K K K S E R A D L I A Y L K D A T A K
Q A E G F S Y T E A N K N K G I T W G E E T L M E Y L E N P K K Y I P G T K M I F A G I K K K A E R A D L I A Y L K D A T S K
Q A A G F S Y T D A N K N K G I T W G E D T L M E Y L E N P K K Y I P G T K M I F A G I K K K G E R Q D L I A Y L K S A C S K
Q A E G Y S Y T D A N K S K G I V W N N D T L M E Y L E N P K K Y I P G T K M I F A G I K K K G E R Q D L V A Y L K S A T S -
Q A A G F A Y T N A N K A K G I T W Q D D T L F E Y L E N P K K Y I P G T K M I F A G L K K P N E R G D L I A Y L K S A T K -
Q A P G F S Y S N A N K A K G I T W G D D T L F E Y L E N P K K Y I P G T K M V F A G L K K A N E R A D L I A Y L K E S T K -
T T A G Y S Y S A A N K N K A V E W E E N T L Y D Y L L N P K K Y I P G T K M V F P G L K K P Q D R A D L I A Y L K K A T S S
S V D G Y A Y T D A N K Q K G I T W D E N T L F E Y L E N P K K Y I P G T K M A F G G L K K D K D R N D I I T F M K E A T A -
Q A Q G Y S Y T D A N I K K N V L W D E N N M S E Y L T N P K K Y I P G T K M A F G G L K K E K D R N D L I T Y L K K A C E -
Q A Q G Y S Y T D A N K R A G V E W A E P T M S D Y L E N P K K Y I P G T K M A F G G L K K A K D R N D L V T Y M L E A S K -
H K D N Y A Y S E S Y K A K G L T W T E A N L A A Y V K N P K A F V L E S K M T F K - L T K D D E I E N V I A Y L K T L K - -
                    T E M                              K S G D P K A K
```

```
Q A P G Y S Y T A A N K N K G I T W G E D T L M E Y L E N P K K Y I P G T K M I F V G I K K K E E R A D L I A Y L K K A T N E
Q A P G F S Y T D A N K N K G I T W G E D T L M E Y L E N P K K Y I P G T K M I F A G I K K K G E R A D L I A Y L K K A T N E
Q A P G F S Y T D A N K N K G I T W G E E T L M E Y L E N P K K Y I P G T K M I F A G I K K K G E R A D L I A Y L K K A T N E
Q A P G F S Y T D A N K N K G I T W G E E T L M E Y L E N P K K Y I P G T K M I F A G I K K K G E R E D L I A Y L K K A T N E
Q A   G F S Y T D A N K N K G I T W G E D T L M E Y L E N P K K Y I P G T K M I F A G I K K K G E R A D L I A Y L K   A T S K
Q A E G F S Y T D A N K N K G I T W G E D T L M E Y L E N P K K Y I P G T K M I F A G I K K K   E R A D L I A Y L K D A T S K
Q A E G F S Y T D A N K N K G I T W G E D T L M E Y L E N P K K Y I P G T K M I F A G I K K K S E R A D L I A Y L K D A T S K
Q A   G F S Y T D A N K N K G I T W G E D T L M E Y L E N P K K Y I P G T K M I F A G I K K K G E R   D L I A Y L K S A T S K
Q A   G Y S Y T D A N K N K G I T W G E D T L M E Y L E N P K K Y I P G T K M I F A G I K K K G E R   D L I A Y L K S A T S -
Q A A G Y S Y T   A N K N K G I T W G E D T L F E Y L E N P K K Y I P G T K M   F A G L K K       E R A D L I A Y L K   A T   -
Q A A G F S Y T N A N K A K G I T W G D D T L F E Y L E N P K K Y I P G T K M   F A G L K K   N E R A D L I A Y L K   A T K -
Q A A G Y S Y T   A N K N K G     W   E N T L F E Y L E N P K K Y I P G T K M   F   G L K K       D R A D L I A Y L K   A T   -
Q A   G Y S Y T D A N K     K G     W D E N T L F E Y L E N P K K Y I P G T K M A F G G L K K       K D R N D L I T Y   K E A T   -
Q A Q G Y S Y T D A N K   K G V   W D E N T M S E Y L E N P K K Y I P G T K M A F G G L K K A K D R N D L I T Y   K E A   -
```

the ends of sequences except in the case of the bacterium, where there are internal differences in length. The amino acid positions are numbered according to the wheat sequence, which has 112 amino acids. At 20 positions (*color*) the same amino acid is found in all the sequences, and the degree of identity is far greater among related species. These observed sequences constitute the raw data

that are fed into the computer. The computer determines the ancestral sequences that can best account for the relations among observed sequences. These ancestral sequences establish the nodes: locations at which the branches of the phylogenetic tree diverge. Node 1, for example, is the ancestor of the primates, Node 2 is the mammalian ancestor and Node 10 is the vertebrate ancestor.

and to have preserved through natural selection—such similar structures. On the other hand, gradual evolution from a common ancestor through millions of generations provides a convincing explanation for both the similarities and the differences among present-day cytochrome sequences.

The evolutionary process is made possible by mutations: errors in the copying and passing along of genetic material from generation to generation. The most frequently accepted mutation within a gene is the exchange of a single nucleotide for another, which may yield a protein that has one amino acid changed.

A second kind of error is the duplication of a portion of a gene. This can yield an elongated gene or, often, two almost complete copies of the original genetic material that proceed to mutate independently. Finally, nucleotides can be deleted or inserted, resulting in a protein of altered length.

c

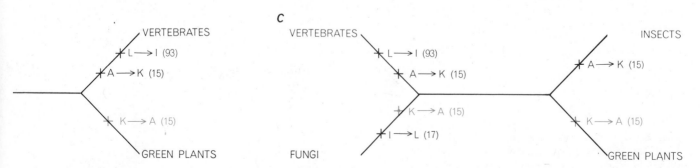

place between them. This reasoning, together with similar reasoning from evidence at Position 17 and Position 93 (*see text*), suggests a certain topological relation of lineages (*a*). It includes three muta-

tions. There are two other possible configurations (*b, c*) but they each require four mutations, two of which have alternative forms (*color*). The first topology is therefore the most probable one.

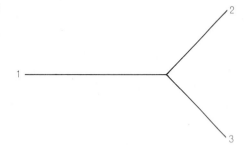

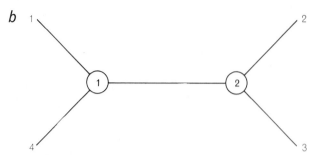

COMPUTER PROGRAM 2 builds an approximate topology by beginning with three observed sequences, which can only be related by a simple three-branch topology (*a*). It then adds a fourth sequence to each of the three original branches in turn, establish-

Over billions of years many such errors have occurred in individual organisms. A few have been selected as beneficial and perpetuated in the species; most have been deleterious and have been eliminated. One pressure against their selection is the biochemical conservatism that results from the interdependence of the various cell components. A protein must automatically fold into a precise three-dimensional shape when it is synthesized, and the shape is predetermined by the sequence. Each protein becomes adapted to performing a particular function in which it must interact with other components, whether through its chemical action, through the complexes it forms, through the rate at which its reactions proceed or through its structural properties. Moreover, all these capabilities must be little disturbed by changes or extremes in the environment.

Under such circumstances as these, most changes in protein sequence—even if they are advantageous for a particular function—are likely to disturb so many other interactions as to be almost always deleterious on balance. So severe are these constraints that an identical sequence of each protein is found in most individuals of a species, and a given sequence may be predominant in a species for several million years. Occasionally a minor variant may be tolerated, persist and eventually become preferable because of a change in other cell components or in the external environment. In other cases a rarely occurring error may immediately prove to be beneficial. Sometimes the environmental circumstances are so strained or the beneficial error is so profound that two separate populations or even two species develop. Subsequent changes arise independently in the two separate groups.

The degree of difference among present-day species and the order of their derivation from common ancestors are commonly represented by a phylogenetic tree. It is possible to derive such a tree from protein-sequence data. The

basic method is to infer from observed sequences the ancestral sequences of the proteins from which they diverged, and thus to establish a series of nodes that define the connections of twigs to branches and of branches to limbs. Then all the observed and reconstructed sequences and the topology, or order of branching, that connects them are considered at once, and the configuration that is most consistent with the known characteristics of the mutational process is chosen. Within the limitations of the small quantity of data available so far, a tree constructed in this way has the same topology as trees that have been derived from conventional morphological or other biological considerations. When the structure of a large number of proteins has been worked out, there will be enough evidence to establish the order of divergence of the major living groups of organisms and even a relative time scale for these divergences. The detailed nature and order of acceptance of mutations that occurred in the distant past may then become clear.

Each point on a phylogenetic tree derived from protein sequences represents a definite time, a particular species and a predominant protein structure for the

individuals of the species. For any such tree there is a "point of earliest time"; radiating from this point, time increases along all branches, with protein sequences from present-day organisms at the ends of the branches. The location of the point of earliest time—the connection to the trunk of the tree—cannot be inferred directly from the sequences; it must be estimated from other evidence.

To illustrate some of the general considerations in building a phylogenetic tree let us consider just three amino acid positions in the cytochrome *c* sequences (excluding the bacterial one). It is clear, first of all, that biologically similar organisms tend to have the same amino acid in a given position. In Position 15 the plants all have the amino acid alanine (*A*), whereas the animals have lysine (*K*). In Position 17 the fungi (*Neurospora*, yeast and *Candida*) have leucine (*L*), whereas the wheat and most animal sequences have isoleucine (*I*); only a fish (the tuna) has threonine (*T*). In Position 93 the insects and plants have leucine, whereas the vertebrates have isoleucine.

Changes arise so seldom that an observed change almost always reflects a mutation in a single ancestral organism.

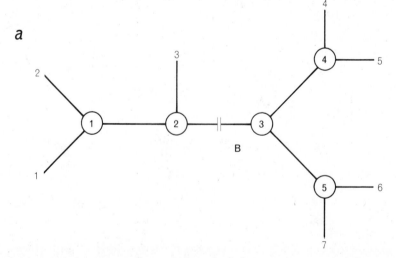

PROGRAM 3 improves on the topology established by Program 2 by trying alternative configurations. It does this by cutting each branch of the tree and grafting the resulting pieces

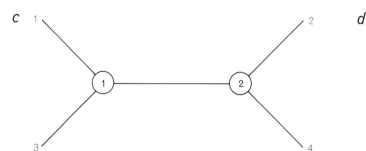

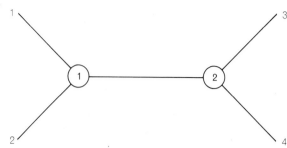

ing three possible topologies for a four-branch tree (*b, c, d*). Program 1 establishes the sequences for the nodes (*numbered circles*)

and evaluates each topology. The best one is accepted. In this way all the sequences are added until a complete tree has been formed.

The evidence at Position 15 favors the hypothesis that there was a single mutation in a single lineage connecting the animal group with the plant group. The mutation at Position 17 indicates a single lineage between fungi and the other species; the one at Position 93 indicates a single lineage between vertebrates on the one hand and insects and plants on the other. Taken together, these pieces of evidence yield a topology that accommodates all the information from the three sites and requires that only three changes occurred in three ancestral organisms. There are two other possible topologies, but they require that at least four changes must have taken place [*see bottom illustration on pages 112 and 113*]. Since changes in sequence are so rare, we assume that the first configuration is the one most likely to be correct.

It is necessary, of course, to consider all the evidence, not just that found at three amino acid positions. Evidence

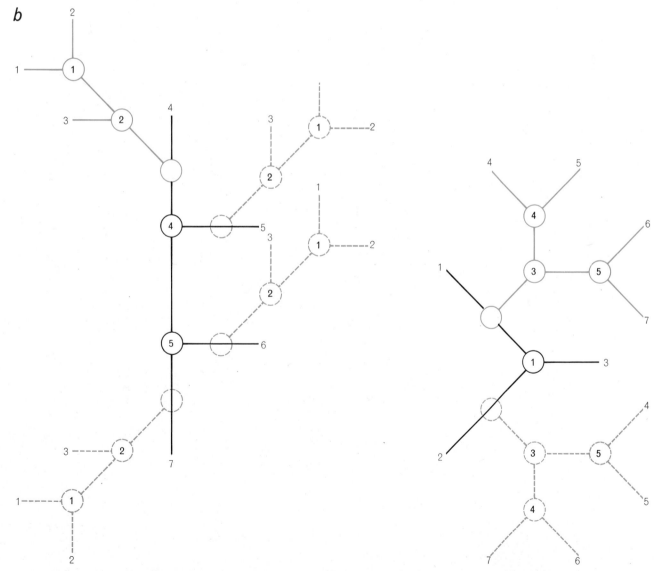

in different ways. In this example a small tree (*a*) is divided into parts *A* and *B*. Four new topologies are created (*b*) when *A* is

grafted to *B* at four points (*color*). Two new structures result (*c*) when *B* is grafted to *A*. The procedure is repeated for each branch.

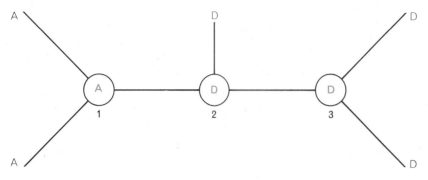

PROGRAM 1, which evaluates topologies, infers the ancestral sequences at each node. It does this one amino acid at a time. In this tree the amino acids at a certain position in five observed sequences are shown (*black letters*). From this information the amino acids at that position in ancestral sequences at Node 1, Node 2 and Node 3 are inferred (*color*).

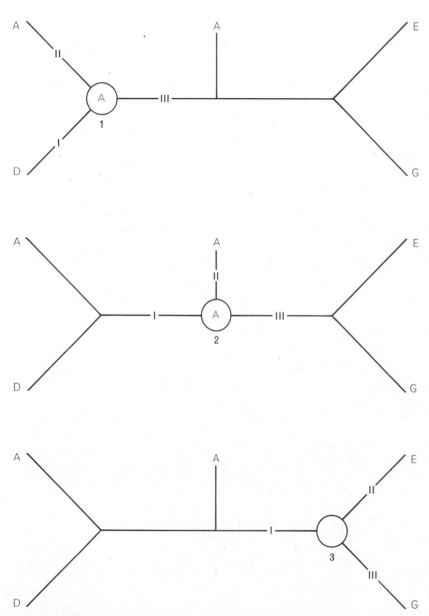

PROCEDURE followed by the computer is to make a list, for each node, of the amino acids on each branch (*Roman numerals*). The amino acid that is on more lists than any other one is assigned to the node. Here the lists would read, for Node 1, D, A and AEG (*top*); for Node 2, DA, A and EG (*middle*); for Node 3, DAA, E and G (*bottom*). Amino acid A appears on two lists for Node 1 and so it is assigned to the node. For the same reason it is assigned to Node 2. No amino acid is clearly the best for Node 3 and so it is left blank.

from a number of other positions confirms the choice of the first topology described above, but occasionally there is conflicting evidence. For example, at Position 74 there is evidence that wheat and *Candida* are in one group that is connected by a single lineage to all other species. Since this is contrary to the weight of all the other evidence, we must assume that in this position there were two distinct mutations, in two different groups, that by coincidence yielded the same amino acid.

In constructing a phylogenetic tree the quantity of data to be considered is so large and objectivity is so essential that processing the information is clearly an appropriate task for a computer. Our approach is to make an approximation of the topology and then try a large number of small changes in order to find the best possible tree. We have developed three computer programs to do this. Program 1 evaluates a topology. It does this, as I shall explain in more detail below, by first determining the ancestral sequences at all the nodes in a given topology and then counting the total number of amino acid changes that must have occurred in order to derive all the present-day sequences from the ancestral ones. The lower the number, the better the topology is assumed to be. The other two programs use Program 1 to build an approximate tree and then to improve it.

Program 2 starts with three observed present-day sequences, which can only have a simple three-branch topology. It then adds a fourth sequence to each of the three branches in turn [*see top illustration on preceding two pages*] and applies Program 1 to evaluate each resulting topology. The best one is chosen. Then a fifth sequence is added, and then, one at a time, the rest of the sequences. Since each placement is decided without regard to the sequences to be located later, at least one wrong decision is very likely to be made, producing a tree that is almost but not quite the best one. Program 3 is therefore applied to shift each of the branches to other parts of the tree, thus testing all likely alternative configurations. This can be done systematically by cutting each branch or group of branches and grafting it to every other branch or limb of the tree [*see bottom illustration on preceding two pages*]. Again the resulting topologies are evaluated by Program 1, and the best one is finally chosen.

Program 2 and Program 3 are straightforward in logic, although they were intricate programming problems. Our

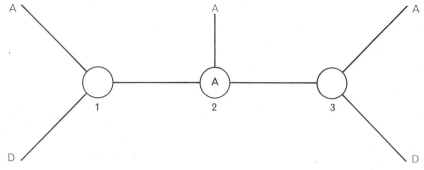

FURTHER STEPS are required of Program 1 to avoid leaving unnecessary blanks. In this example the basic procedure would leave Node 1 and Node 3 blank (because at both nodes A and D would each appear on two lists. The program therefore examines the first nodal assignments and, if at least two of the three positions adjacent to a blank node (including another node) have the same amino acid, assigns that amino acid to the blank node. Thus A is assigned to Node 1 and Node 3. Ultimately each node must agree with two of its neighbors.

major decisions were made in designing Program 1, which evaluates the topologies proposed by the other two programs.

Program 1 begins by making inferences about the ancestral sequences to be assigned to the nodes. It does this by considering, one at a time, the amino acid positions along the chain. Where only one amino acid is found in all the observed sequences, almost certainly it was present in all the ancestors at all times. In less clear-cut situations a number of reasonable conjectures can be made regarding the ancestral sequences. Consider the case of the amino acids at a certain position in five sequences connected by a definite topology [see top illustration on opposite page]. What was the ancestral amino acid at that position in the three nodal sequences? At Node 1 it is most likely to have been A, and at Node 2 and Node 3 it is most likely to have been D. There was, then, one mutation between Node 2 and Node 1. Any other assignment of amino acids would require two or more mutations.

Let us now see how the computer handles such a problem in practice. The computer must treat all possible topologies, not just one particular case. For this purpose any tree can be thought of as being made up entirely of nodes connecting three branches [see bottom illustration on opposite page]. More complex branching simply involves two or more such nodes with zero distance between them. Each of the three branches connects the node either with an observed sequence or with another node and, through it, ultimately with two or more other observed sequences. The computer makes a list of the amino acids that lie on each branch. For example, the lists for Node 1 would show D on Branch I, A on Branch II and A, E and G on Branch III. Then the amino acid that is found on more lists than any other is assigned to each node. If no single first choice exists, the position is left blank. By this procedure A would be selected for Node 1 and Node 2; Node 3 would be left blank.

In a number of situations this simple program gives an equivocal assignment when it need not [see top illustration on this page]. The procedure I have described would assign blanks to Node 1 and Node 3 although it is intuitively clear that the choice of A for all three nodes is best, necessitating two independent mutations from A to D. Any other choice would require at least three mutations, a less likely history.

We therefore added further steps to enable the computer to fill in unnecessary blanks. The first assignment of nodal amino acids is examined. If at least two of the three positions adjacent to a blank node contain the same amino acid, that amino acid can be inferred also for the blank node. This second assignment may supply the information required to fill in other blanks, and so the procedure is repeated until no more changes occur. Finally any node that does not have the same amino acid as two of its neighbors is changed to a blank. The entire process yields a definite assignment of ancestral amino acids wherever one choice is clearly preferable and leaves blanks where there is reasonable doubt. By applying these procedures to each position the program eventually spells out all the ancestral sequences.

The nodal sequences for cytochrome c are displayed along with the observed sequences [see top illustration on pages 112 and 113]. The very small number of blanks indicates how few of the positions remain doubtful. These computed ancestral sequences, incidentally may take

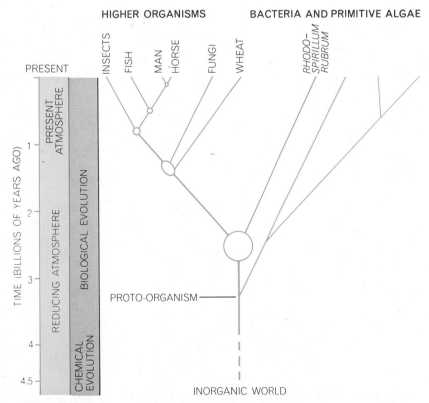

CYTOCHROME C TREE (dark color) is redrawn on an absolute time scale and in the context of earth history. The bacterial branch has been added and the "point of earliest time" is taken to be equidistant from the bacterial and the other present-day sequences (see text). The size of each node reflects the degree of uncertainty in determining its position.

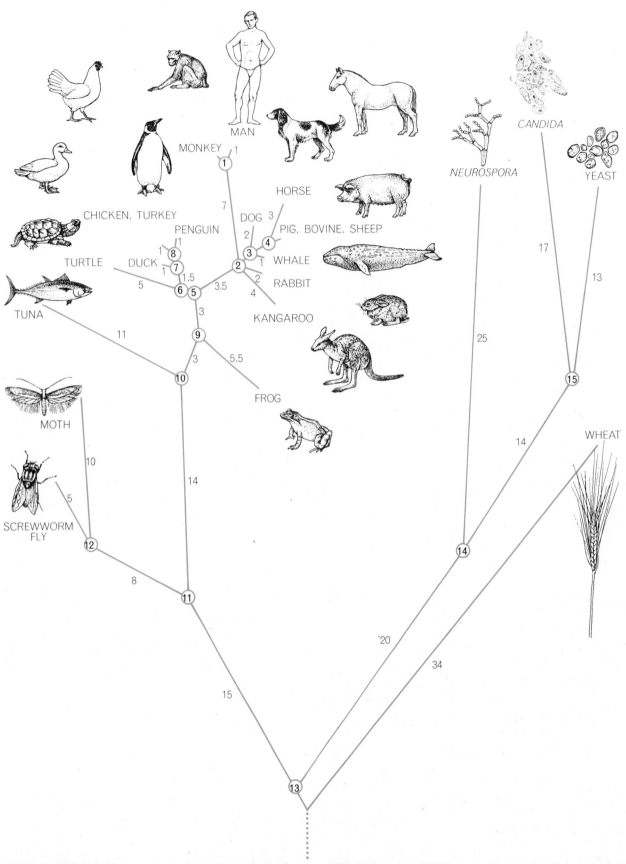

MAN

MONKEY

HORSE

DOG

PIG, BOVINE, SHEEP

CHICKEN, TURKEY

PENGUIN

WHALE

DUCK

RABBIT

TURTLE

KANGAROO

TUNA

FROG

MOTH

SCREWWORM
FLY

NEUROSPORA

CANDIDA

YEAST

WHEAT

PHYLOGENETIC TREE showing the derivation of present-day organisms was constructed on the basis of a computer analysis of homologous proteins of cytochrome *c*, a complex substance that is found in similar versions in different species. The sequence of the amino acids that constitute the homologous protein chains is slightly different in each of the organisms shown at the ends of the branches (*see illustration on pages 112 and 113*). Analysis of the differences reveals the ancestral relations that dictate the topology of the tree. The computer programs determine the sequences of the unknown ancestral proteins shown at the nodes of the tree (*numbered circles*) and compute the number of mutations that must have taken place along the way (*numbers on branches*).

on real meaning in view of the increasing possibility of synthesizing proteins in the laboratory. As investigators succeed in duplicating the sequences we may learn a great deal about the chemical capabilities of ancient organisms.

Once the ancestral sequences have been established, the amino acid changes along each branch of the tree are totaled. (Even when a position is left blank, it is possible to determine the number of changes that must have taken place there.) The sum, representing the total number of changes on the tree, is the final score for that tree. In this way each of the alternative topologies proposed by Program 2 and Program 3 is evaluated in turn.

To make the best topology into the best phylogenetic tree one needs a measure of branch length. We use the number of mutations between nodes. The figures for the observed amino acid changes, however, understate the actual number of mutations because mutations can be superimposed: in a long enough time interval, for example, an *A* might change to a *D* and the *D* to an *S* and the *S* to an *A*, eradicating the evidence of change. We correct for superimposed mutations by applying factors based on the known probability that any amino

acid will change to any other given amino acid. That provides our unit of branch length: accepted point mutations per 100 amino acid positions (PAM's). Now it is **possible to draw the tree** [*see illustration, opposite*]. The major groups fall clearly into the topology shown, but some of the details are still uncertain. It is hard to establish the exact sequence of events in the short interval during which the lines to the kangaroo, the rabbit, the ungulates and the primates diverged. For some divergences, such as the one to the dog and the gray whale, the topology depends on a single amino acid position, and there is perhaps one chance in five or 10 that the branching point is incorrect by one unit. In time other protein sequences from these animals should clear up the uncertainties.

It remains only to establish a time scale for the tree and, by establishing a point of earliest time, to relate the history of cytochrome *c* to geologic time. Our impression is that a protein such as cytochrome *c*, once its function is well established, is subject to about the same risk of mutation in a given time interval no matter what species it is in. It may well turn out that this is not true—that the risk varies in major groups and that occasionally a species may undergo a large change. For the time being, how-

ever, we assume that the mutation rate is constant over long intervals, and we define a time scale in terms of the number of mutations.

The bacterial sequence provides information with which to establish the point of earliest time. Because the *Rhodospirillum* sequence is so different from the other sequences it is not shown on the cytochrome *c* tree. There is evidence for its placement, however. At Position 13 and Position 29 *Rhodospirillum* and wheat are different from all the other sequences but are like each other. This indicates that the bacterium should be attached to the wheat branch. Then the fungi and the animals must have diverged from each other after the line to higher plants diverged from the bacteria. To allow for its many differences, the bacterial branch must be very long—about 95 PAM's. That being the case, the point of earliest time must be well back on the bacterial branch.

Now it is possible to redraw the cytochrome tree in simplified form with a time scale in years. The translation from PAM's to years is derived from geological evidence, the best of which dates the divergence of the lines to the bony fishes and the mammals at about 400 million years ago. The cytochrome tree puts that divergence at 11.5 PAM's, on the aver-

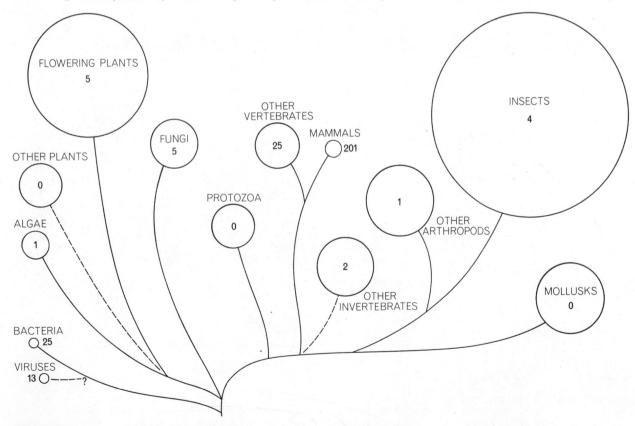

SEQUENCE DATA are accumulating rapidly. The numbers indicate how many sequences of 30 amino acid positions or more have been determined in each biological group; the area of the circles is proportional to the number of species described in each group (except in bacteria and viruses, where species are not clearly defined). Data from a wide range of groups are needed for paleogenetics.

age. Therefore 11.5 PAM's corresponds to 400 million years. Now the major nodes can be plotted on the basis of the average number of mutations on the branches above each node. We assume, moreover, that the point of earliest time— the connection to the trunk of the tree— is equidistant from the bacterial and the other present-day sequences. Thus from one family of related proteins we can estimate the temporal relations for an extensive tree of life [see bottom illustration on page 117].

The tree is shown in the context of current theory from geological and other biochemical considerations. Elso S. Barghoorn of Harvard University has reported fossil evidence that at least two kinds of organism existed more than three billion years ago, one resembling a bacterium and the other a blue-green alga.

Their common ancestor, the "proto-organism," must already have had a complex cell chemistry; it contained many related proteins presumably descended by evolutionary processes from fewer ancestral proteins. Before the time of the proto-organism there was an era of chemical evolution through which life emerged from an inorganic world [see "Chemical Fossils," by Geoffrey Eglinton and Melvin Calvin; beginning on page 79].

The rate of change over geologic time varies greatly from one protein to another. Cytochrome c protein is the most slowly changing one that has been studied so far in a wide variety of organisms; it appears to have changed at the rate of about 30 PAM's every billion years. The comparable figure for insulin is 40; for the enzyme glyceraldehyde-3-phosphate dehydrogenase, 20; for histones (proteins

that are bound to DNA) only .6 PAM. On the other hand, hemoglobin has undergone about 120 PAM's per billion years, ribonuclease 300 and the fibrinopeptides, which are involved in the chemistry of blood clotting, 900. It seems likely that some of the slowly changing proteins will provide the best information on long-term evolution because they have undergone fewer superimposed mutations. Proteins that change more rapidly will provide higher resolution for sorting out closely related species.

Each protein sequence that is established, each evolutionary mechanism that is illuminated, each major innovation in phylogenetic history that is revealed will improve our understanding of the history of life. Surely insight into the biochemistry of man will be obtained from better understanding of his origins.

V

EXTRATERRESTRIAL LIFE

V EXTRATERRESTRIAL LIFE

INTRODUCTION

Ed Deevey, Jr., once wrote, "There is a unique and nearly ubiquitous compound with the empirical formula $H_{2960}O_{1480}C_{1480}N_{16}P_{1.8}S$ called living matter." This is certainly true for our planet. Is it true for other planets of the solar system? For planets of other stars?

The arguments presented throughout have their roots in the assumption of mediocrity. Life is the outcome of those forces that shape and evolve terrestrial planets. Implicit within this argument is the notion that the process is inevitable. Do we fear the darkness of being alone so much that we must construct elaborate hypotheses as an escape? Or do our hypotheses contain so much truth (details aside for now) that in cold hard fact we are as natural an occurrence as any rock?

We can test hypotheses in the laboratory only to a limited extent. We have seen through all these articles, which are only the tip of the iceberg, that widely disparate observations lead to a comforting concordance of conclusions. Still, we could be the outcome of unique events, which appear likely one by one but which when taken together may be highly improbable. For example, the Gold garbage hypothesis suggests that our stem cells, from which all life arose, were bacterial garbage dumped by some ancient astronauts as they explored the primitive Earth. This simple chance event serves to "explain" why Earth life bears a universal biochemistry. But what was the origin of these ancient astronauts? The basic problem has really been shelved rather than solved.

The ultimate test of the hypothesis that the origin of life is inevitable will be to examine life that has arisen independently. If this life has the same kind of genetic and protein-synthesizing apparatus, if it has a cellular basis, if it uses L-amino acids and D-sugars, if it has the forms we do, then and only then can we vindicate our theory.

The NASA *Viking 1* mission searched for life on Mars. Norman Horowitz reports the outcome—no life was found. Does this mean that our theory is incorrect? Not really. Mars is a very different planet from the Earth. Since it is so much less massive, it probably never generated enough internal heat to have the intense early volcanic activity that the Earth and Venus experienced. This means that liquid water, carbon dioxide, nitrogen, and other volatiles remain locked within the planet, unavailable for the formation of oceans and an atmosphere. Under such conditions life cannot originate. Admittedly, Mars does possess a slight atmosphere and some frozen water. There is even a hint of some slight past volcanic activity, but not enough. However, Mars was our best chance within the solar system to find other life. Now we must turn to the stars.

Su-Shu Huang surveys the vastness of the stars within our galaxy and other galaxies, computing the probability that terrestrial planets within the proper temperature zone and of the proper size should be a common occurrence. Thus our theory for the origin of life is supported by an untestable computation, for the stars are not within our present reach.

The uncertainty of our origin remains as a significant driving force for research. What more fascinating legacy could we pass to the next generation?

Suggested Further Reading

Bracewell, Ronald N. 1975. *The Galactic Club: Intelligent Life in Outer Space.* W. H. Freeman and Company, San Francisco.

Ponnamperuma, Cyril, and A. G. W. Cameron. 1974. *Interstellar Communication.* Houghton Mifflin, Boston.

Shklovskii, I. S., and Carl Sagan. 1966. *Intelligent Life in the Universe.* Holden-Day, San Francisco.

Viking 1: Early Results. 1976. National Aeronautics and Space Administration SP-408. Scientific and Technical Information Exchange Office, Washington, D.C.

12

The Search for Life on Mars

by Norman H. Horowitz
November 1977

The Viking landers have completed their biological experiments. The experiments did not detect life process, but they did reveal much of interest about the chemistry of the surface of the planet

Is there life on Mars? The question is an interesting and legitimate scientific one, quite unrelated to the fact that generations of science-fiction writers have populated Mars with creatures of their imagination. Of all the extraterrestrial bodies in the solar system Mars is the one most like the earth, and it is by far the most plausible habitat for extraterrestrial life in the solar system. For that reason a major objective of the Viking mission to Mars was to search for evidences of life.

The two Viking spacecraft were launched from Cape Canaveral in the summer of 1975. Each spacecraft consisted of an orbiter and an attached lander. When the spacecraft arrived at Mars in July and August of 1976, each was put in a predetermined orbit around the planet, and the search for a landing place began. Cameras aboard the orbiters were the principal source of information on which the choice of the landing sites was based; important data also came from infrared sensors on the orbiters and from radar observatories on the earth. The sole consideration in the final selection of the sites was the safety of the spacecraft. It would be a mistake to suppose, however, that the sites were therefore without biological interest. Biological criteria dominated the initial decisions as to the latitude at which each spacecraft would land. Once the latitudes had been chosen there was relatively little difference between sites at different longitudes.

On command from the earth each lander separated from its orbiter. With the help of its retroengines and parachute it dropped to the surface of Mars. Both orbiters continued to circle the planet, operating their own scientific instruments and relaying to the earth data transmitted from the landers. Both landings were in the northern hemisphere of Mars, and the Martian season was summer. (Mars has seasons like those on the earth, but each season lasts approximately twice as long. The Martian year is 687 Martian days; each Martian day, named a sol by the Viking team to distinguish it from a terrestrial day, is 24

hours 39 minutes long.) On July 20, 1976, the *Viking 1* lander came to rest in the Chryse Planitia region of Mars, some 23 degrees north of the equator. Six weeks later the *Viking 2* lander settled down in the Utopia Planitia region, some 48 degrees north of the equator. In longitude the two landers are separated by almost exactly 180 degrees, thus placing them on opposite sides of the planet. Since the instrumentation of the two landers is identical, the difference in their landing sites is the only distinction between them.

The first biologically significant task carried out by each lander was the analysis of the Martian atmosphere. Life is based on the chemistry of light elements, notably carbon, hydrogen, oxygen and nitrogen. To be suitable as an abode of life a planet must have those elements in its atmosphere. Spectroscopic observations from the earth and from spacecraft that had flown past Mars in previous years had already shown that carbon dioxide was the principal component of the Martian atmosphere. Small quantities of carbon monoxide, oxygen and water vapor had also been detected. Nitrogen had not been detected in any form, however, and atmospheric theory suggested that Mars had lost most of its nitrogen in the past.

Each Viking lander analyzed the atmosphere by means of two mass spectrometers. One spectrometer, operating during the descent to the surface, sampled and analyzed the atmospheric gases every five seconds. The second spectrometer operated on the ground. The results showed that the atmosphere near the ground was approximately 95 percent carbon dioxide, 2.5 percent nitrogen and 1.5 percent argon, and that it also held traces of oxygen, carbon monoxide, neon, krypton and xenon. At both landing sites the atmospheric pressure was 7.5 millibars. (The atmospheric pressure at sea level on the earth is 1,013 millibars.)

Since the Viking spacecraft revealed that nitrogen is indeed present in the Martian atmosphere, we can say that

the elements necessary for life are available on Mars. Missing from the list of gases, however, is one critically important compound: water vapor. Although earlier measurements had shown that traces of water vapor are present in the Martian atmosphere, the quantity varies with season and place. The Viking orbiters carried out a survey of water vapor over the entire planet with infrared spectrometers. The results showed that the highest concentration of atmospheric water vapor was at the edge of the north polar cap (the summertime hemisphere), and that the concentration fell off toward the south (the opposite of what is found on the earth). In the polar region the amount of water vapor in the atmosphere would form a film only a tenth of a millimeter thick if all of it were to be condensed on the planet's surface. At the landing sites the concentration of water vapor ranged between 10 and 30 percent of the concentration at the pole.

These numbers put into quantitative terms a long-known fact about Mars: It is a very dry place. Mars has ice at its poles, but nowhere on its surface are there oceans or lakes or any other bodies of liquid water. The absence of liquid water is related to the dryness of the atmosphere through a fundamental law of physical chemistry: the phase rule. The phase rule states that for liquid water to exist on the surface of a planet the pressure of the water vapor in the atmosphere must at some times and in some places be at least 6.1 millibars. The Viking measurements imply that the vapor pressure of water at the surface of Mars in the northern hemisphere is at most .05 millibar, even if all the water vapor is concentrated in the lower atmosphere. At that low pressure liquid water cannot remain in the liquid phase; depending on the temperature, it must either freeze or evaporate. By the same token raindrops cannot form in the Martian atmosphere and ice cannot melt on the Martian surface.

The extreme dryness presents a difficult problem for any Martian biology. Liquid water is essential for life on the

HOROWITZ

earth. All terrestrial species have high and apparently irreducible requirements for water; none could live on Mars. If there is life on Mars, it must operate on a different principle as far as water is concerned. If Mars had a more favorable environment in the past, how-

ever, and if the planet did not dry up too fast, species may have had time to evolve and adapt to present conditions. Pictures made by the *Mariner 9* spacecraft, which went into orbit around Mars in 1971, suggested that Mars may indeed have had running water on its

surface in the past. The pictures from the Viking orbiters have confirmed that impression. The evidence consists of channels in the Martian desert that resemble dry riverbeds. There seems to be little doubt that the channels were carved by rapidly flowing liquid, and

LANDING SITE OF VIKING 1 LANDER on Mars was photographed from the spacecraft in February to document the digging of the series of trenches seen at the lower right. Soil samples from the trenches were delivered to instruments in the spacecraft to be tested for their chemical composition and for signs of life. The orange tinge of sky shows that a great amount of dust is suspended in atmosphere.

LANDING SITE OF VIKING 2 LANDER is a field of boulders superficially similar to the landing site of the *Viking 1* lander. The camera aboard the lander made this picture early in the afternoon of September 24, 1976, three weeks after spacecraft had landed and before samples of soil were taken. Horizon is tilted because lander is not quite level. Objects in foreground are instruments on spacecraft.

there is widespread agreement that the most probable liquid is water.

If liquid water once existed on Mars, could life have arisen on the planet? If the life evolved to meet changing conditions, could it exist there still? There is no way to settle these questions by deductive reasoning or even by experimentation in laboratories on the earth. They can be answered only by the direct exploration of Mars, and that is what the Viking spacecraft did.

Five different types of instrument on each Viking lander were involved in the search for evidences of life: two cameras for photographing the landscape, a combined gas chromatograph and mass spectrometer for analyzing the surface for organic material and three instruments designed to detect the metabolic activities of any microorganisms that might be present in the soil. In this brief account I shall not be able to mention the names of the many scientists, engineers and managers whose joint efforts made all the Viking projects possible. They work in universities, industrial laboratories and the National Aeronautics and Space Administration and its field centers. Their names are recorded in the growing technical literature dealing with this historic mission.

Each of the Viking landers carried two cameras of the facsimile type, which built up a picture of the scene by scanning it in a series of narrow strips. Such cameras make pictures slowly, but they are rugged and versatile. Their resolution was moderately high: a few millimeters at a distance of 1.5 meters. They produced pictures in black and white, in color and in stereo. The two cameras on each lander could between them survey the entire horizon around the spacecraft.

As life-seeking tools cameras have inherent advantages and disadvantages. Their chief advantage lies in the fact that a picture contains a large amount of information. In principle it would be possible to prove unequivocally the existence of life on Mars with a single photograph. For example, if a line of trees were visible on the horizon or if footprints appeared on the ground in front of the spacecraft one morning, there would be no room for doubt that there is life on Mars. Another advantage lies in the fact that pictorial evidence is independent of all assumptions about the chemistry and physiology of Martian organisms. The organisms need not respond in certain ways to certain substances or treatments in order to be recognized. The cameras could identify, say, a mushroom made of titanium as a form of life if one were to sprout up from under a rock in the course of the mission. Of course, reliance on pictorial evidence rests on its own set of assumptions about the morphology of living things. The most obvious disadvantage of the camera as a life-seeking instrument is the fact that an entire world of life can exist below the camera's limit of resolution.

Of all the results of the Viking mission the wonderful photographs of the Martian desert at the two landing sites are the most impressive. The photographs have been eagerly scanned by alert and hopeful eyes, but no investigator has yet seen anything suggesting a living form.

The next step was to analyze the soil for any organic constituents. Among the elements carbon is unique in the number, variety and complexity of the compounds it can form. The special properties of carbon that enable it to form large and complex molecules arise from the basic structure of the carbon atom. That structure enables the carbon atom to form four strong bonds with other atoms, including other carbon atoms. The molecules thus formed are very stable at ordinary temperatures, so stable, in fact, that there seems to be no limit to the size they can attain. The connection between life and organic chemistry (that is, the chemistry of carbon) rests on the fact that the attributes by which we identify living things—their capacity to replicate themselves, to repair themselves, to evolve and to adapt—originate in properties that are unique to large organic molecules. It is the highly complex information-rich proteins and nucleic acids that endow all the living things we know, even "simple" ones such as bacteria and viruses, with their essential nature. No other element, including that favorite of science-fiction writers, silicon, has the capacity carbon has to form large and complex structures that are so stable. It is no accident that even though silicon is far more abundant than carbon on the earth, it has only minor and nonessential roles in biochemistry. Biochemistry is largely a chemistry of carbon.

Such fundamental facts lead to the conclusion that wherever life arises in the universe it will most likely be based on carbon chemistry. That view has been strengthened by the discovery of organic compounds of biological interest in meteorites and in clouds of dust in interstellar space. Although these compounds are nonbiological in origin, they are closely related to the amino acids and the nucleotides that are the respective building blocks of proteins and of nucleic acids. The fact that they are

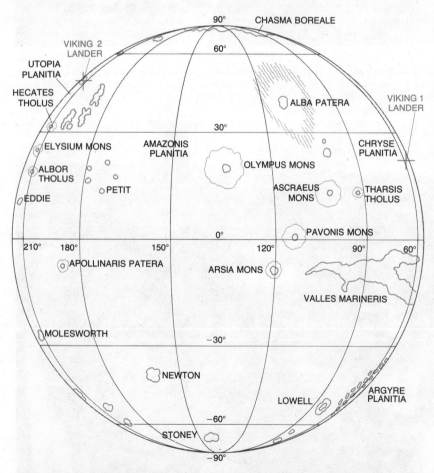

LOCATIONS OF THE TWO VIKING LANDERS are indicated on this map of Mars, which shows some of the major geological features of the planet. The two spacecraft are on opposite sides of the planet, some 4,600 miles apart. Both are in the northern hemisphere at sites selected partly for their possible biological interest. *Viking 1* lander is in Chryse Planitia region at a latitude of 23 degrees; *Viking 2* lander is in Utopia Planitia region at a latitude of 48 degrees.

formed in settings remote from the earth implies that carbon chemistry gives rise to familiar organic compounds throughout the universe. This fact in turn suggests that life elsewhere in the universe will be based on an organic chemistry similar to our own, although not necessarily identical with it.

Such considerations led to the decision to include an organic-analysis experiment aboard the Viking landers. The instrument used in the experiment was the mass spectrometer that had analyzed the atmosphere combined with a gas chromatograph and a pyrolysis furnace. A sample of the Martian soil was first heated in the furnace through a series of steps up to a temperature of 500 degrees Celsius. Any volatile materials released were passed through the gas chromatograph. Since each of the different compounds has a different molecular weight, composition and polarity, among other properties, it passed through the columns of the gas chromatograph at a unique rate, and so the compounds were separated from one another. As each compound emerged from the chromatographic column it was directed into the mass spectrometer for identification. Since essentially all organic matter is cracked, or decomposed, into smaller fragments at 500 degrees C., the method is capable of detecting organic compounds that have a wide range of molecular weights.

Two soil samples were analyzed at each landing site. The only organic compounds detected were traces of cleaning solvents known to have been present in the apparatus. The fact that the solvents were detected shows the instruments were functioning properly. The heated samples gave off carbon dioxide and a small amount of water vapor; nothing else was found.

This result is surprising and weighs heavily against the existence of biological processes on Mars. The combined gas chromatograph and mass spectrometer aboard each Viking lander is a sensitive instrument, capable of detecting organic compounds at a concentration of a few parts per billion, a level that is between 100 and 1,000 times below their concentration in desert soils on the earth. Even if there is no life on Mars, it has been supposed the fall of meteorites onto the Martian surface would have brought enough organic matter to the planet to have been detected. Because Mars is near the asteroid belt, from which meteorites originate, it is believed to receive a much larger number of meteorite impacts than either the earth or the moon. Indeed, a question that was frequently discussed before the Viking spacecraft were launched was whether or not it would be possible to distinguish biological organic matter on Mars from the meteoritic organic matter that was expected to be

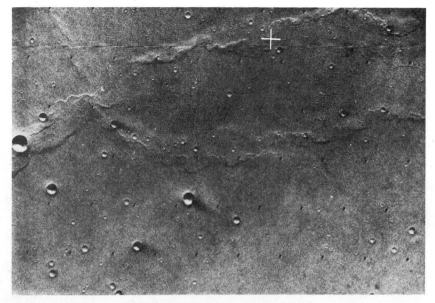

CHRYSE REGION was photographed from 1,555 kilometers above the Martian surface by the *Viking 1* orbiter on July 17, 1976, showing a recently determined accurate landing site of the *Viking 1* lander (*cross*). The landing area, in the western part of Chryse Planitia, is a smooth plain with many small impact craters dotting the surface. The wrinkle ridges to the west (*top*) seem to be similar to volcanic ridges found on smooth lava floors of maria on the moon.

present. The absence of organic matter at the parts-per-billion level, however, suggests that on Mars organic compounds are actively destroyed, probably by the strong ultraviolet radiation from the sun.

The other experiments aboard the Viking landers searched not just for organic matter in the soil but for living organisms. On the earth microorganisms such as bacteria, yeasts and molds are the hardiest of all species. There are few places on the earth where microbial forms do not live; they are the last survivors in environments of extreme temperature and aridity. The reasons for their hardiness are interesting but need not detain us here. Suffice it to say that if there is life on Mars, the chance of detecting it would be maximized by searching for microorganisms in the Martian soil.

Each Viking lander carried three instruments designed to detect the metabolic activities of soil microorganisms. First, the gas-exchange experiment was designed to detect changes in the composition of the atmosphere caused by

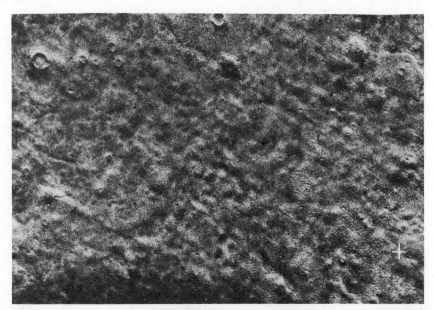

UTOPIA REGION was photographed from 3,360 kilometers above the surface of Mars by the *Viking 2* orbiter on August 16, 1976, as the spacecraft was surveying the planet for a landing site for the *Viking 2* lander. The area is rough, apparently blanketed by dunes. Site at which the *Viking 2* lander touched down is indicated by cross at far right (north) edge of photograph.

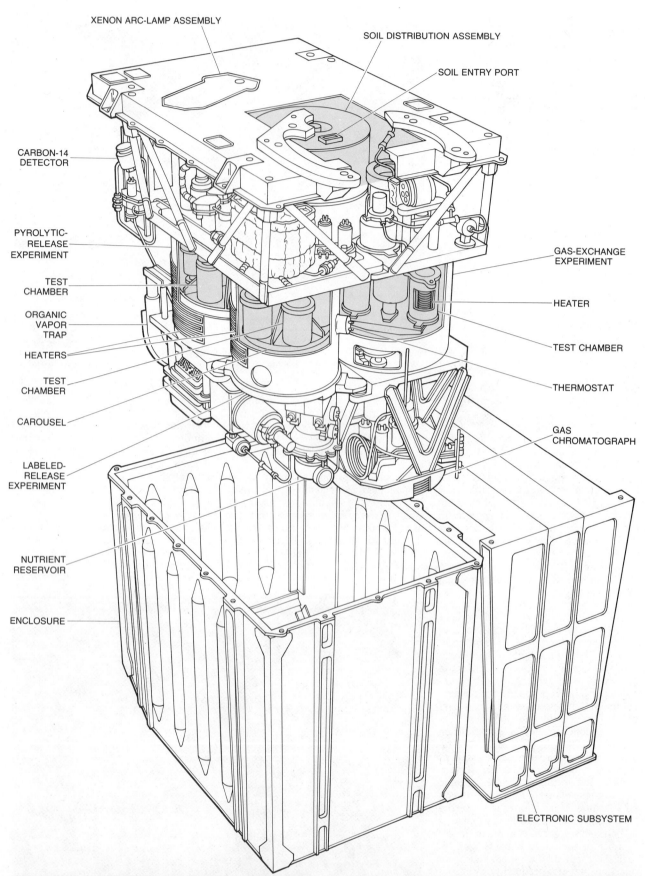

XENON ARC-LAMP ASSEMBLY

SOIL DISTRIBUTION ASSEMBLY

SOIL ENTRY PORT

CARBON-14 DETECTOR

PYROLYTIC-RELEASE EXPERIMENT

TEST CHAMBER

ORGANIC VAPOR TRAP

HEATERS

TEST CHAMBER

CAROUSEL

LABELED-RELEASE EXPERIMENT

NUTRIENT RESERVOIR

ENCLOSURE

GAS-EXCHANGE EXPERIMENT

HEATER

TEST CHAMBER

THERMOSTAT

GAS CHROMATOGRAPH

ELECTRONIC SUBSYSTEM

BIOLOGICAL LABORATORY aboard both Viking spacecraft occupies a volume of only one cubic foot. The three biological experiments were the gas-exchange experiment (*right*), the labeled-release experiment (*bottom left*) and the pyrolytic-release experiment (*top left*). Each experiment, shown cut away, had several test chambers on a carousel so that the experiment could test several samples of Martian soil. The soil was dumped into an entry port at the top of the laboratory, where it fell into a hopper. For each sample of soil one test chamber of each experiment was rotated under the hopper in order to receive a portion of the sample. All together approximately half a dozen samples of soil were tested at each landing site. The experiments were completed in April. Results are given on following pages.

microbial metabolism. Second, the labeled-release experiment was designed to detect decomposition of organic compounds by soil microbes when they were fed with a nutrient. Third, the pyrolytic-release experiment was designed to detect the synthesis of organic matter in Martian soil from gases in the atmosphere by either photosynthetic or non-photosynthetic processes. All three experiments analyzed portions of each sample of Martian soil.

All the experiments detected chemical changes of one kind or another in the soil. All the experiments are now completed, and some of the changes they observed suggest biological processes. There has been much discussion both within the team of Viking investigators and outside it as to the best way to interpret the findings. Are the changes due to biological responses or are they just chemical reactions we would like to believe are biological? Indeed, since life is a form of chemistry, how can the two be told apart?

One way to decide whether or not a process is biological is to test its sensitivity to heat. Living structures are highly organized and fragile, and they are destroyed by temperatures that leave many chemical reactions unaffected. A process that is insensitive to heat is thus likely to be a nonliving chemical reaction, but a process that is sensitive to it could be either living or nonliving.

The decision as to whether a heat-sensitive process is biological or not must be based on additional evidence. In the end, however, the judgment is based on Occam's razor: the traditional principle that the hypothesis most likely to be correct is the one that accounts for the maximum number of observations with the minimum number of assumptions.

The gas-exchange experiment and the labeled-release experiment were frankly terrestrial in orientation. In both experiments a nutrient medium composed of an aqueous solution of organic compounds was mixed with a sample of Martian soil. Since liquid water cannot exist on Mars, the experiments could not be conducted under Martian conditions; the test chambers had to be heated to prevent the water from freezing and pressurized to prevent it from boiling. Both experiments were based on the universal property of terrestrial organisms to evolve gas as they metabolize food. If a sample of soil from the earth is moistened with a nutrient solution, the microorganisms in the soil take up the nutrients and convert them partly into more microorganisms (that is, the population of microorganisms grows) and partly into various by-products, including gases. Among the gases given off in microbial metabolism are carbon dioxide, methane, nitrogen, hydrogen and hydrogen sulfide. On the earth gases evolved by one species of organisms are

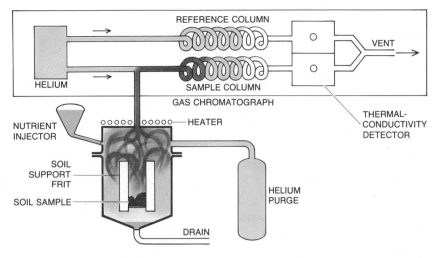

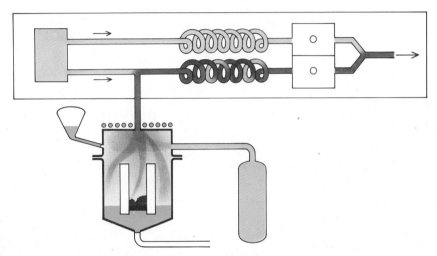

GAS-EXCHANGE EXPERIMENT tested the Martian soil to see if there were any microorganisms in it that took in atmospheric gases and nutrients and gave off gaseous by-products. The experiment proceeded in two stages. In the first stage a small volume of a complex nutrient solution (*dark color*) was injected into the test chamber in such a way that it humidified the chamber without wetting the soil (*top*). The gases evolved (*light gray*) were flushed into a gas chromatograph with a stream of helium (*light color*), where they were analyzed for organic compounds and compared with results of reference analysis run as a standard. In second stage of experiment a large volume of nutrient was poured into chamber to wet the soil (*bottom*).

eventually consumed by other species of organisms. In that way the light elements at the earth's surface are continually cycled through the biosphere and the atmosphere.

In the gas-exchange experiment a complex nutrient solution was added to a sample of Martian soil in a closed chamber, and the gases were analyzed periodically by means of a gas chromatograph. The experiment proceeded in two stages. In the first stage a small volume of the nutrient solution was introduced into the soil chamber in such a way that it humidified the chamber without actually wetting the soil, and the resulting gases were analyzed several times. In the second stage a large volume of the nutrient was poured into the chamber, saturating the soil. With the soil now in direct contact with the medium the main

part of the experiment began. The soil was incubated for nearly seven months, so that whatever microorganisms might be in the sample had enough time to signal their presence by producing or consuming gases. During the period of incubation the atmosphere in the chamber was periodically analyzed.

The findings of the first stage of the experiment were both surprising and simple. Immediately after the soil sample was humidified carbon dioxide and oxygen were rapidly released. The release of the gases ceased soon after it had begun but not before the pressure in the chamber had risen measurably. At the Chryse site in a period of little more than one sol the quantity of carbon dioxide in the incubation chamber of the *Viking 1* lander increased by a factor of five and the quantity of oxygen increased by a factor of 200. At the Utopia site the increases were less, but they were still considerable.

The rapidity and the brevity of the response recorded by both landers clearly suggested that the process observed was a chemical reaction, not a biological one. The appearance of the carbon dioxide is readily explained. Carbon dioxide gas would be expected to be adsorbed on the surface of the dry Martian soil; if the soil was exposed to a very humid atmosphere, the gas would be displaced by water vapor. The appearance of the oxygen is more complex. The production of so much oxygen seems to require an oxygen-generating chemical reaction, not just a physical liberation of pre-existing gas. It is likely that the oxygen was released when the water vapor decomposed an oxygen-rich compound such as a peroxide. Peroxides are known to decompose if they are exposed to water in the presence of iron compounds, and according to the X-ray fluorescence spectrometer aboard each Viking lander, the Martian soil is 13 percent iron.

At both landing sites the second phase of the gas-exchange experiment was anticlimactic. When the soil sample was saturated with the nutrient medium and incubated, carbon dioxide continued to be released. The production of the carbon dioxide gradually tapered off, however, and the oxygen gradually

disappeared. The slow increase in the amount of carbon dioxide was probably a continuation of the reaction in the humid stage of the experiment. The disappearance of the oxygen also can be easily explained: one of the ingredients of the nutrient medium was ascorbic acid, which combines readily with oxygen. And so after seven months it became clear that everything of interest had happened in the humid stage of the experiment, before the soil came in contact with the nutrient! What the gas-exchange experiment detected was not metabolism but the chemical interaction of the Martian surface material with water vapor at a pressure that has not been reached on Mars for many millions of years.

The labeled-release experiment differed from the gas-exchange experiment in several ways. The nutrient medium employed was a simpler one containing only a few cosmically abundant organic compounds such as formic acid ($HCOOH$) and the amino acid glycine (NH_2CH_2COOH). All the compounds were labeled with atoms of the radioactive isotope carbon 14. The labeled-release instrument was designed to detect radioactive gases, principally carbon dioxide, released when the nutrient medium was added to a sample of soil. The number of radioactive disintegrations in gases can be counted quite efficiently, so

that the labeled-release experiment is faster and more sensitive than the gas-exchange experiment in detecting microbial activity in terrestrial soil. The labeled-release experiment's sequence of operations did not include a humid stage as such, but it attempted to accomplish the same end by injecting a volume of nutrient medium that was insufficient to wet the entire soil sample but sufficient to humidify the chamber. If the experiment worked on Mars as planned, subsequent injections of the medium, which were controlled by commands sent from the earth, brought the medium into contact with some soil that had previously been wetted and with other soil that had been humidified but not wetted.

As in the gas-exchange experiment, immediately after the nutrient medium was added to the soil in the labeled-release experiment, gas surged into the chamber. The release of gas tapered off soon after the first sol. The gas, undoubtedly carbon dioxide, was radioactive, showing that it had been formed from the radioactive compounds of the medium and not from compounds in the Martian soil. Nonradioactive gases, which also must have formed when the aqueous medium came in contact with the soil, were not detectable in the experiment.

The production of radioactive carbon dioxide in the labeled-release experiment is understandable in the light of the evidence from the gas-exchange experiment suggesting that the surface material of Mars contains peroxides. Formic acid, one of the compounds of the labeled-release nutrient medium, is oxidized with particular ease: if a molecule of formic acid ($HCOOH$) reacts with one of hydrogen peroxide (H_2O_2), it will form a molecule of carbon dioxide (CO_2) and two molecules of water ($2H_2O$). The amount of radioactive carbon dioxide given off in the labeled-release experiment was only slightly less than what would have been expected if all the formic acid in the medium had been oxidized in this way.

If the source of the oxygen released in the humid stage of the gas-exchange experiment was indeed peroxides in the soil decomposed by water vapor, then in the labeled-release experiment all the peroxides should also have been decomposed by the first injection of nutrient. Thus the next injection should have evolved no additional radioactive gas in spite of the fact that part of the sample presumably had not yet been wetted by the medium. That proved to be the case. When a second volume of medium was injected into the chamber, the amount of the gas in the chamber was not increased; indeed, it decreased. The decrease is explained by the fact that carbon dioxide is quite soluble in water; when fresh nutrient medium was added

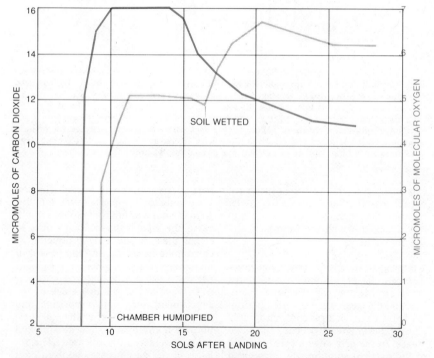

RESULTS OF THE GAS-EXCHANGE EXPERIMENT, according to data of Vance I. Oyama of the Ames Research Center of the National Aeronautics and Space Administration, showed that in the first humid stage of the experiment a large amount of carbon dioxide (black) and molecular oxygen (color) surged into the test chamber. In the second wet stage the amount of carbon dioxide continued to rise at a decreasing rate and then declined. The amount of oxygen, however, quickly fell. It is believed the gases were released by physical and chemical processes, not by biological ones. One micromole is a millionth of a mole, where one mole is the amount of a substance that has a weight in grams equal to its molecular weight. Oxygen curve is displaced to the left by one sol so that the curves do not overlap. A sol is one Martian day.

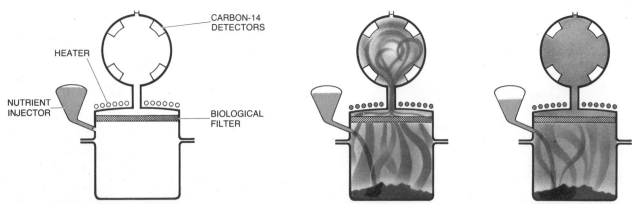

HEATER

NUTRIENT
INJECTOR

CARBON-14
DETECTORS

BIOLOGICAL
FILTER

LABELED-RELEASE EXPERIMENT on the Viking landers tested the Martian soil for microorganisms that could metabolize simple organic, or carbon, compounds. The nutrient medium was composed of several organic compounds that are widely abundant in the universe. The compounds were labeled with radioactive carbon. If microorganisms exist in the Martian soil, they might consume the labeled nutrient and give off radioactive gases (particularly carbon dioxide), which would be detected by the carbon-14 counters. Before the soil was tested the background level of radiation was measured (*left*).

The soil was dumped into the test chamber, injected with a small amount of medium (*middle*) and incubated for up to 11 sols. The amount of nutrient in this first injection was planned to wet only part of the soil but to humidify the entire chamber. A subsequent injection of the nutrient (*right*), controlled by signals from the earth, thus brought the medium into contact with soil that had already been wetted and with other soil that had been humidified but not wetted. If the labeled-release experiment worked on Mars as planned, its results would serve as a check on the results of the gas-exchange experiment.

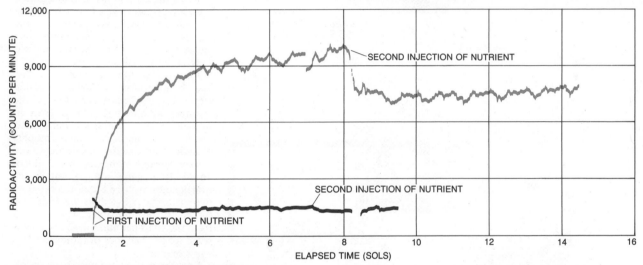

SECOND INJECTION OF NUTRIENT

SECOND INJECTION OF NUTRIENT

FIRST INJECTION OF NUTRIENT

RESULTS OF THE LABELED-RELEASE EXPERIMENT for the first sample of soil analyzed at the Chryse site (*curve in color*) were indeed consistent with the results of the gas-exchange experiment, according to data from Gilbert V. Levin and Patricia A. Straat of Biospherics Inc. Immediately after the first injection of the nutrient, radioactive gases surged into the chamber. The radioactivity was measured at 16-minute intervals throughout the experiment except for the first two hours after the first injection, when measurements were made every four minutes. After the second injection the amount of gas in the chamber dropped, then remained at a nearly constant level until the end of the experiment. In order to test the sensitivity of reaction to heat a second portion of the soil sample was sterilized at a temperature of 160 degrees Celsius for three hours and the experiment was repeated (*curve in black*). The reaction was abolished. Although such behavior is consistent with a biological process, it is more likely that experiment again detected only a chemical reaction.

to the chamber, it absorbed some of the carbon dioxide in the head space above the sample.

This result was obtained with all the samples tested by the labeled-release experiment at both Viking sites. In that respect the results of the labeled-release experiment did not parallel the results of the gas-exchange experiment. At both sites both experiments tested soil gathered from the ground's exposed surface; at the Utopia site the experiments also tested soil gathered from under a rock. Although the labeled-release experiment found essentially no difference in the amount of gas released by any of the samples, the gas-exchange experiment recorded about three-fourths as much

carbon dioxide from the surface samples at the Utopia site as it had from the surface samples at the Chryse site, and it recorded even less carbon dioxide from the sample from under the rock.

The gas-exchange experiment also recorded less oxygen from the samples from the Utopia region, but the interference of the ascorbic acid in the complex nutrient medium of that experiment makes it difficult to quantify the difference. In every case, however, the gas-exchange experiment detected considerably more gas than the labeled-release experiment did with portions of the same sample. Those results, however, do not contradict the thesis that the production of oxygen detected by the gas-exchange

experiment and the production of radioactive carbon dioxide detected by the labeled-release experiment are simply different measurements of the same surface chemistry. The gas-exchange experiment measures the total amount of oxidant in the surface; the labeled-release experiment measures only a fraction of it.

The labeled-release experiment also tested the stability of the reaction to heat. When the soil was preheated to 160 degrees C. for three hours before incubation, the reaction was abolished. When it was heated to 46 degrees for the same length of time, the magnitude of the reaction was reduced by about half. These results have been regarded by

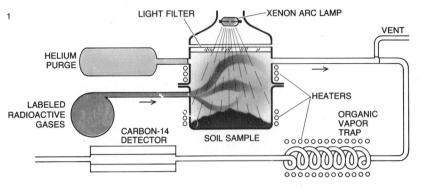

1

LIGHT FILTER — XENON ARC LAMP

VENT

HELIUM
PURGE

LABELED
RADIOACTIVE
GASES

HEATERS

CARBON-14
DETECTOR

ORGANIC
VAPOR
TRAP

SOIL SAMPLE

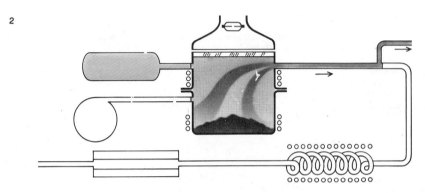

2

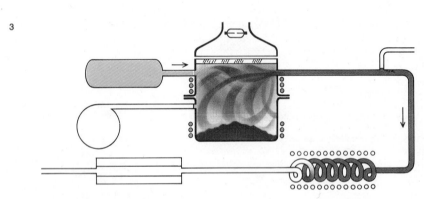

3

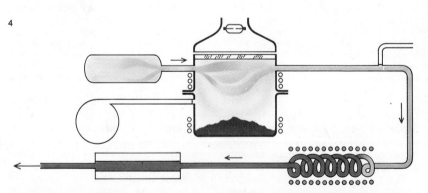

4

PYROLYTIC-RELEASE EXPERIMENT tested the Martian soil (*dark gray*) to see if there were microorganisms in it that would create organic compounds out of atmospheric gases by either a photosynthetic process or a nonphotosynthetic process. A sample of soil was sealed into a chamber along with some Martian atmosphere and a small amount of radioactive carbon dioxide and carbon monoxide (*light gray*). A xenon arc lamp irradiated the soil with simulated Martian sunlight (*1*). After five days lamp was turned off and the atmosphere was removed from the chamber (*2*). Soil was heated to a temperature high enough to pyrolyze (decompose) into small volatile fragments any radioactive organic compounds produced. Fragments (*light gray*) were swept out of the chamber (*3*) by a stream of helium (*light color*) into a column designed to trap organic molecules but pass carbon dioxide and carbon monoxide. In column trapped radioactive organic molecules were released by raising column's temperature; the molecules were oxidized to form carbon dioxide, which was carried into a radiation counter (*4*).

some as evidence in favor of the hypothesis that the reaction is biological. The results are of course consistent with such a hypothesis, but they are also consistent with a chemical oxidation in which the oxidizing agent is destroyed or evaporated at relatively low temperatures. A variety of both inorganic peroxides and organic peroxides could probably have produced the same results.

The third microbiological experiment, the pyrolytic-release experiment, differed from the gas-exchange and labeled-release experiments in two respects. First, it attempted to measure the synthesis of organic matter from atmospheric gases rather than its decomposition. Second, it was designed to operate under the conditions of pressure, temperature and atmospheric composition that actually obtain on Mars, since those are the conditions under which any form of Martian life must exist. In practice the conditions in the chamber were a reasonably good approximation of Martian conditions except for the temperature, which stayed warmer than the outside temperature because of heat sources within the spacecraft.

A sample of Martian soil was sealed in a chamber along with some Martian atmosphere. A quartz window in the chamber admitted simulated Martian sunlight from a xenon arc lamp. Into this Martian microcosm small amounts of radioactive carbon dioxide and radioactive carbon monoxide were introduced. Both gases are present in the Martian atmosphere but not in radioactive form. After five days the lamp was turned off, the atmosphere was removed from the chamber and the soil was analyzed for the presence of radioactive organic matter.

First the soil was heated in the pyrolysis furnace to a temperature high enough to crack any organic compounds into small volatile fragments. The fragments were swept out of the chamber by a stream of helium and passed through a column that was designed to trap organic molecules but allow carbon dioxide and carbon monoxide to pass through. The radioactive organic molecules were thus transferred from the soil to the column and at the same time were separated from any remaining gases of the incubation atmosphere. The organic molecules were released from the column by raising the column's temperature. Simultaneously the radioactive organic molecules were decomposed into radioactive carbon dioxide by copper oxide in the column. The carbon dioxide was then carried by the stream of helium into a radiation counter. If organic compounds had been synthesized in the soil, they would be detected as radioactive carbon dioxide; if no organic compounds had been synthesized, no radioactive carbon dioxide would have been formed.

Surprisingly, seven of the nine pyro-lytic-release tests executed on Mars gave positive results. The two negative results were obtained at the Utopia site, but a third sample tested at Utopia was positive. This third sample was actually incubated in the dark, implying that light may not be required for the re-action. The amount of carbon fixed in the soil by the experiment was small: enough to furnish organic matter for be-tween 100 and 1,000 bacterial cells. The quantity is so small, in fact, that it could not have been detected by the organic-analysis experiment. The quantity is nonetheless significant; it was surprising that in such a strongly oxidizing envi-ronment even a small amount of organic material could be fixed in the soil.

Even more significant, the pyrolytic-release instrument had been rigor-ously designed to eliminate nonbiologi-cal sources of organic compounds. Dur-ing the development of the experiment it had been found that in the presence of short-wavelength ultraviolet radiation, carbon monoxide spontaneously com-bined with water vapor to form organ-ic molecules on glass, quartz and soil surfaces in the experimental chamber. In order to avoid those reactions and the confusion they would have caused, the short-wavelength ultraviolet was fil-tered out of the radiation allowed to en-ter the incubation chamber. To receive positive results from the soil on Mars in spite of that precaution was startling.

Nevertheless, it appears that the find-ings of the pyrolytic-release experiment must also be interpreted nonbiological-ly. The reason is that the reaction detect-ed was less sensitive to heat than one would expect of a biological process. In two of the nine pyrolytic-release experi-ments performed on Mars the soil sam-ple was heated before the radioactive gases were injected and the incubation was begun. In one case the sample was held at 175 degrees C. for three hours and in the other it was held at 90 degrees for nearly two hours. The effect of the higher temperature was to reduce the reaction by almost 90 percent but not to abolish it. The effect of the lower tem-perature was nil. When it is recalled that the temperature at the surface of Mars at the two landing sites does not rise above zero degrees C. at any time, and that the temperature below the surface is even lower, it becomes difficult to reconcile the results with a biologi-cal source. Any organisms living in the Martian soil should have been killed by those temperatures.

On the other hand, it is not easy to point to a nonbiological explanation for the positive results. Investigations into the problem are now under way in ter-restrial laboratories with synthetic Mar-tian soils formulated on the basis of the data from the inorganic analyses carried out by the Viking landers. The solution

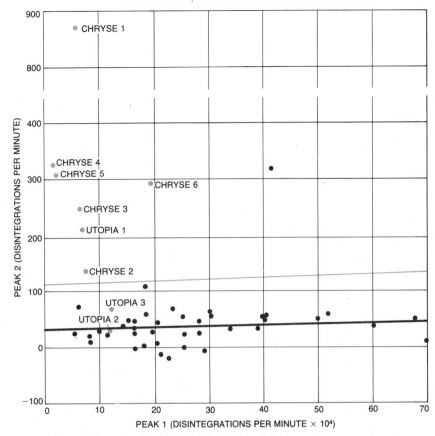

RESULTS OF PYROLYTIC-RELEASE EXPERIMENT are shown for all the samples tested on Mars (*dots in color*). The axis labeled Peak 1 shows how much radioactivity in the form of carbon dioxide and carbon monoxide passed through the column during the pyrolysis of each sample. The axis labeled Peak 2 shows how much radioactivity, representing newly synthesized organic matter, remained attached to the column in each case. Each dot in color is labeled and numbered according to the site at which it was tested and which experiment it represents. (For example, "Chryse 1" means the result of the first experiment at Chryse, "Utopia 1" means the result of the first experiment at Utopia, and so on.) The dots in black are the data ob-tained from tests of sterilized soil samples in a duplicate of the Viking pyrolytic-release instru-ment on the earth. The black line drawn through those points represents the best fit to the points. The colored line above the black line is a statistically significant dividing line; any point lying above the colored line is a positive result. The single black point above the colored line is believed to be due to a technical error in performing that particular test. Seven of the nine pyrolytic-release experiments performed on Mars, however, yielded firmly positive results.

to the puzzle will probably also explain why the organic-analysis experiment de-tected no organic material in the Mar-tian surface. Until the mystery of the results from the pyrolytic-release exper-iment is solved, a biological explanation will continue to be a remote possibility.

Even though some ambiguities re-main, there is little doubt about the meaning of the observations of the Vi-king landers: At least those areas on Mars examined by the two spacecraft are not habitats of life. Possibly the same conclusion applies to the entire planet, but that is an intricate problem that cannot yet be addressed. The most surprising finding of the life-seeking ex-periments is the extraordinary chemical reactivity of the Martian soil: its oxidiz-ing capacity, its lack of organic matter down to the level of several parts per billion and its capacity to fix atmospher-ic carbon (presumably into organic mol-ecules) at a still lower level. It seems Mars has a photochemically activated

surface that, due to the low temperature and the absence of water, is maintained in a state far from chemical equilibrium.

These conclusions drawn from the re-sults of the life-seeking experiments on the Viking landers are undeniably disap-pointing. The discovery of life would have been much more interesting, to say the least. There are doubtless some who, unwilling to accept the notion of a life-less Mars, will maintain that the inter-pretation I have given is unproved. They are right. It is impossible to prove that any of the reactions detected by the Vi-king instruments were not biological in origin. It is equally impossible to prove from any result of the Viking experi-ments that the rocks seen at the landing sites are not living organisms that hap-pen to look like rocks. Once one aban-dons Occam's razor the field is open to every fantasy. Centuries of human expe-rience warn us, however, that such an approach is not the way to discover the truth.

13 Life Outside the Solar System

by Su-Shu Huang
April 1960

How many stars in our galaxy may be accompanied by planets capable of supporting intelligent life? The answer requires reflection on the characteristics and life history of stars

The evolution of stars and the evolution of living organisms appear to be completely dissimilar processes. The differences between the two can be explained in terms of how the particles of matter are bound together and how energy is exchanged among them. Indeed, the evolution of living organisms represents one outcome of stellar evolution. It was the steady flow of energy from the sun for four or five billion years that brought about the biological developments on earth, culminating in the emergence of intelligent organisms able to contemplate the whole remarkable story. If the sun had had a different history, life would not have appeared in its immediate vicinity.

Astronomical evidence acquired in recent years indicates that what has happened here is probably not unique. Two decades ago it was thought that the solar system might have originated in a near-collision of the sun and another star, which event supposedly pulled away enough matter from the sun to form the planets. Because such encounters must be rare events, they would give rise to few stars with a company of planets. Today most astronomers believe that a star is formed by the condensation of a cloud of dust and gas [see "The Dust Cloud Hypothesis," by Fred L. Whipple; Scientific American, May, 1948]. This hypothesis much more readily explains the origin of the solar system and is supported by the observation that more than half of the stars in our galaxy are double or multiple systems. In fact,

it now appears that most stars are accompanied by other stars or by planets, though the latter must be so small as to escape sure detection by the present instruments of astronomy. Thus the appearance of life—even the appearance of mind—may be far from unusual events in the universe.

On the other hand, certain critical conditions must be satisfied if life processes are to be initiated and maintained. In the first place the star must shine long enough and steadily enough to permit life to evolve. The star must also be hot enough to warm up a habitable zone deep enough to offer a reasonable chance that a planetary orbit will fall within it. And the planet must ply a stable orbit within this zone. The number of stars that have given rise to life must therefore be considerably smaller than that immediately suggested by the dust-cloud hypothesis.

Taking account of all these factors, what is the probability that man will ever be able to visit the life-bearing planet of another star, or that the earth will receive a visitor from such a planet? Does the possibility justify taking measures now—in advance of a visit —to put existing technology to the task of scanning the sky for signals from intelligent organisms outside the solar system, or for transmitting signals in the hope they may be heard? If so, what sort of signals should be listened for or sent?

A reliable answer to these questions calls for a somewhat closer consid-

eration of the conditions critical for life and an estimate of how often they are likely to be satisfied in the evolution of stars. The first condition is a steady and prolonged flow of energy. How long it takes life to evolve may be judged from the single instance available: Here on earth rational animals evolved from inanimate matter in about three billion years. Biological evolution proceeds by the purely random process of mutation, that is, the unpredictable occurrence of novel chemical processes among the fantastically numerous reactions that constitute life. Since the process is a random one, the laws of probability suggest that the time-scale of evolution on earth should resemble the average time-scale for the development of higher forms of life anywhere. The rate at which mutations occur is of course a variable in the calculation. It is affected by the electromagnetic and corpuscular radiation from the parent star, and the amount of such radiation that reaches the surface of any planet is governed in turn by the magnetic field of the planet, the depth and composition of its atmosphere and other factors. But since mutation itself is a random process, the introduction of a few more variables does not affect the calculation greatly. Furthermore, a higher mutation rate does not necessarily accelerate evolution, because most mutations are harmful. The more frequently they occur, the greater the chance that an individual will suffer an injurious mutation. In order to favor natural selection, mutations should be rare, perhaps

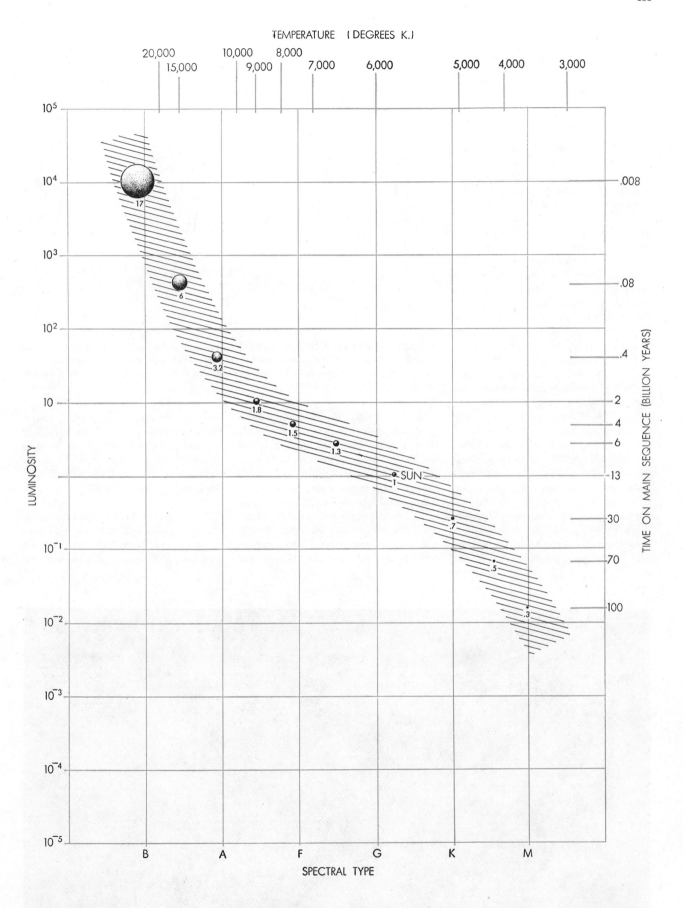

TYPICAL STARS ON THE MAIN SEQUENCE are represented on this diagram. Vertical scale at left indicates luminosity (the sun is unity). Scale at right shows time on the main sequence. The numbers at the top denote temperature in degrees Kelvin; the letters at the bottom, spectral type. Small figures beneath each star give its mass with respect to the mass of the sun.

as rare as in the evolution of life on earth.

It is somewhat easier to estimate the time-scale of stellar evolution. In contrast to the random nature of biological evolution, stellar evolution is governed by the universal law of gravitation and by a relatively small number of thermonuclear reactions. When a star begins to form in a cloud of dust and gas, gravitational attraction among the gas and dust particles causes the cloud to condense until the pressure raises the temperature within it to the point at which the thermonuclear reactions that convert hydrogen to helium begin. The tremendous quantities of energy liberated by these reactions now set up a counterpressure from the center of the star that exactly balances the force of gravitational contraction. In this state of equilibrium the star shines for a much longer time than that required for its condensation out of dust and gas.

The vast majority of the visible stars are in this phase of their evolution. They are called "main-sequence" stars because their luminosity (energy output per unit of time) plotted on a graph against their surface temperature places them in sequence in a narrow band [*illustration on page 135*]. What the chart shows is that the hottest stars are also the most luminous. Luminosity depends upon mass, and so the point at which a star appears on the main sequence depends primarily upon the mass of material incorporated in it during condensation. The hottest stars are designated by the

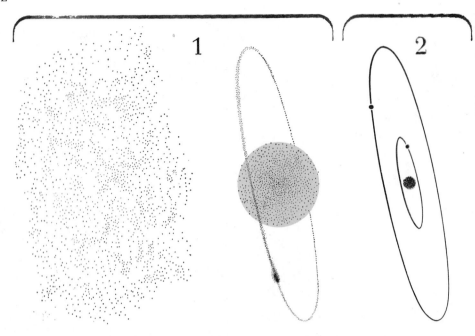

POSSIBLE EVOLUTION OF A STAR 1.2 times the mass of the sun begins with a dust cloud condensing into a star and proto-planets (1), a process taking perhaps 10 million years. Star enters "main sequence" (2) and remains there for approximately eight billion

letter O, followed in descending order by stars classified B, A, F, G, K and M; these classifications are usually called spectral type. The adjectives "early" and "late," which have nothing to do with the age of the star, are often used before the spectral-type designation to denote further relative temperature differences. An early F-type star is hotter than a late one, which in turn has a higher temperature than an early G-type star

such as our sun. The classification is further refined by a number from 0 to 9 written after the letter. Thus the class of early B-type stars includes those from B0 to B4; and the late B-type stars, those from B5 to B9.

The length of time a star remains in the equilibrium state on the main sequence can be calculated from the total mass of hydrogen in its core and the

NOVA IN CONSTELLATION CYGNUS, indicated by pair of white lines, blazed up in 1920. This photograph was made at that time with the 100-inch reflecting telescope at Mount Wilson Observatory. Nova may represent a late stage in the life of a star.

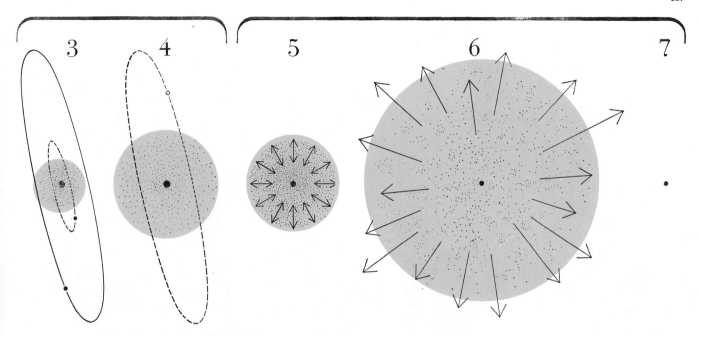

years. Then it expands (3) into red-giant stage (4), first destroying life on its inner planet, then burning up the planets in turn. This period may last about 100 million years. Then star may pulsate in luminosity every few hours (5) for thousands of years, finally exploding into a nova (6) and eventually collapsing into white dwarf (7). Time period for final stages (5, 6 and 7) is not known.

rate at which the hydrogen is consumed. Both factors in the calculation are determined by the position of the star on the main sequence. Luminosity, which is an index of the rate of fuel consumption, increases as the fourth power of the star's mass. The most massive stars thus use up their substance most rapidly and so have the shortest lifetimes in the equilibrium state. In general a star will evolve away from the main sequence when the core in which the hydrogen has been consumed has a mass of about 12 per cent of that of the entire star. With the exhaustion of fuel in the central furnace, gravitational contraction takes over again, heating up the interior until the thermonuclear reaction spreads to outer layers. The star now leaves the main sequence and in a comparatively brief period of time evolves into a red giant or supergiant. Its evolution thereafter cannot presently be predicted in detail. But somehow, perhaps through the rapid loss of mass by ejection, it ends up as a hot, faint, dense object known as a white dwarf. A majority of the intrinsic variable stars, of novae and of nova-like objects are apparently in the stage between red giants and white dwarfs.

In leaving the main sequence a star releases so much energy that it would destroy life on any of its planets. Thus

SAME STAR FADED IN 10 YEARS to the faint object visible between the white lines in this photograph made in 1930 with the same telescope. A nova may explode repeatedly before becoming a white dwarf and eventually turning into a cold, dark body.

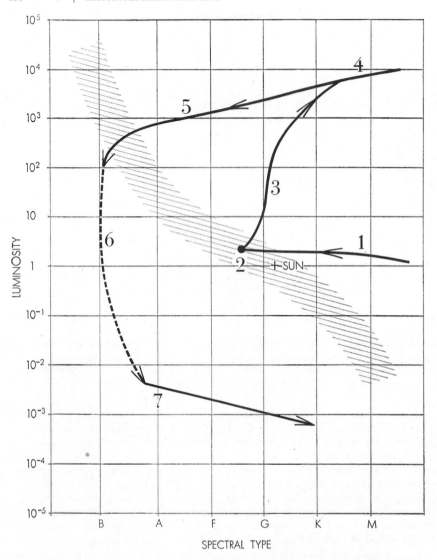

EVOLUTIONARY PATH OF A STAR with a mass 1.2 times that of the sun is traced. The vertical scale is luminosity (the sun is unity); the horizontal scale, spectral type. The main sequence is outlined by the gray hatching. The numbers on the path correspond to the stages of stellar evolution depicted in the illustration at the top of the preceding two pages. At 1 the star is in the stage of gravitational contraction; at 2 it is on the main sequence; at 3 it is expanding; at 4 it is a red giant; at 5 it pulsates. The broken line (6) indicates uncertainty as to the path the star follows in reaching the white-dwarf stage (7).

nor too cold. Thus the chance of finding a planet and intelligent life within the habitable zone of any particular M or even late-K star is quite small. However, since there are 10 times as many M stars as G stars, the total number of life-bearing planets traveling around the M stars may not be negligibly small.

On the basis of their lifetimes and the depth of their habitable zones, stars of the late-F, G and early-K types seem to offer the most favorable environments for life. Approximately 10 per cent of stars fall in these types. Out of the 200 billion stars in the Milky Way, therefore, some 20 billion might foster intelligent life.

This number is reduced considerably upon consideration of the third critical condition: the maintenance of stable planetary orbits. The dust-cloud mechanism that increases the likelihood of planets has also brought most stars into existence as double or multiple systems. It is obvious that the presence of two or more stars in a system will profoundly perturb the orbits of planets in the system. A few double-star systems have members that are so far apart that if they are sufficiently near to us, the stars can be distinguished by a telescope or even by the naked eye. Such systems are called visual binaries. Much more common are the "spectroscopic binaries." In these systems the stars are so close together that even if they are relatively near to us, they cannot be separated by the largest telescope. We can tell that they are double stars only by regular changes in their spectra [*see illustration on page 139*]. These changes also enable us to calculate their orbits.

The orbits of planets in such systems are so complicated that astronomers have been able to work them out for only a few idealized cases. A life-bearing planet would have to travel on an orbit close to one of the stars in a widely separated system, or on an orbit at a large distance from the stars in a close system. In the case of a hypothetical binary composed of two stars with the same luminosity as our sun and revolving around a common center in a nearly circular orbit, a planet would find the thermally habitable zone dynamically stable only if the stars were more than 10 astronomical units or less than .05 astronomical unit apart (an astronomical unit is the distance between the earth and the sun). With a separation between the stars of .5 to 2 astronomical units there is no overlapping of the habitable zone and the dynamically stable

the main-sequence stage of the star's evolution is the only important one so far as life is concerned. The O and early B stars are the most massive, but since they burn much more rapidly than the smaller, less luminous stars, their life on the main sequence lasts only a million to 10 million years. The small M stars, in contrast, remain on the main sequence for more than 100 billion years. None of the early stars in the O, B and A groups has a stable lifetime longer than three billion years; they cannot therefore sustain biological evolution long enough for intelligent organisms to appear on any of their planets. Closer calculation shows that only the stars

farther down the sequence than F4 maintain their equilibrium for a sufficient length of time to bring biological evolution to its culmination.

If time were the only critical condition, then the late-K and M stars would stand the greatest chance of having life-bearing planets. But these stars have low luminosity, and the habitable zones in which planets might travel about them must be quite narrow. A more luminous star can obviously warm up a larger space than a less luminous one [*see illustration on page 142*]. It is like a fire in a field on a cold night: the bigger the fire, the wider the zone around it in which the temperature will be neither too hot

zone. Therefore no inhabitable planets can exist in such systems. Taking everything into consideration, only 1 to 2 per cent of all double and multiple stars may possess inhabitable planets, and perhaps 3 to 5 per cent of all the stars in our galaxy have such planets.

For the present there is no hope of detecting on a photographic plate the existence of a planet of another star. Such planets as may attend even the nearest star are completely lost to view in the brilliance of the star's light. Someday there may be a telescope on a plat-

form in space, free of the interfering effects of the earth's atmosphere. As has been suggested by Nancy G. Roman of the National Aeronautics and Space Administration, the instrument would produce a sharp star image that could be blocked out so that a planet near the

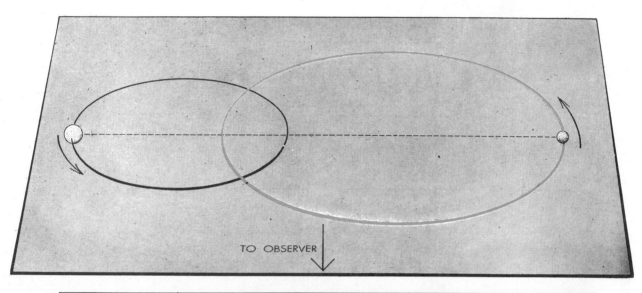

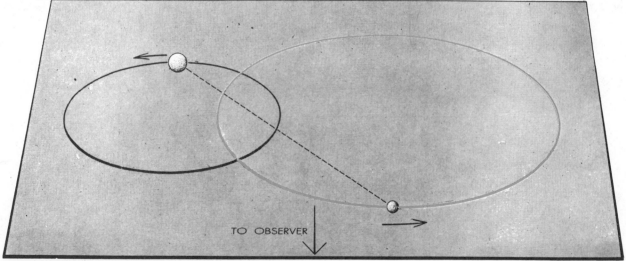

SPECTROSCOPIC DETECTION OF CLOSE BINARIES is depicted in these diagrams and spectra. At top the stars are orbiting around center of mass, the one at left moving toward the earth, the one at right away from the earth. Spectral lines made by star moving toward earth shift toward violet end of the spectrum, while lines from star moving away shift toward red end, splitting spectral lines as seen in upper spectrum below diagrams. In lower diagram the binaries are moving across line of vision as seen from earth. This gives rise to normal single spectral lines seen in the lower spectrum. Spectroscopic binaries are so close together that no telescope can resolve them into separate bodies, and only their spectra enable us to detect them and to calculate their orbital movements.

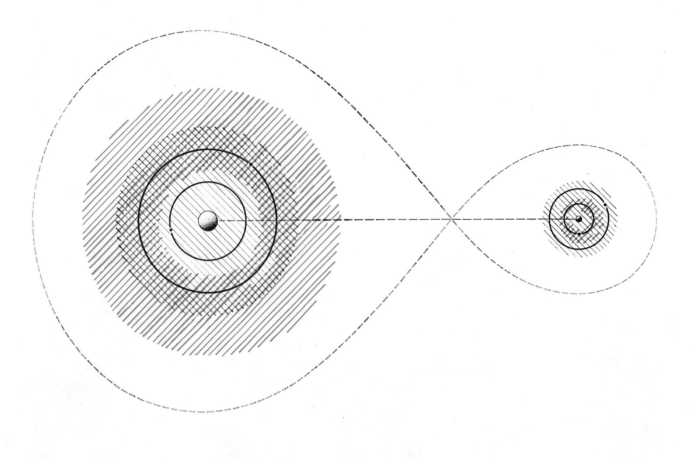

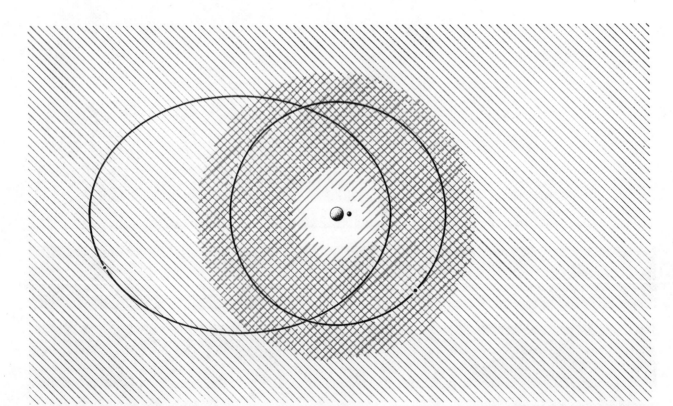

PLANETARY ORBITS AROUND BINARY STARS are depicted here in an idealized manner. The figure eight in the top diagram is simply a mathematical projection from which the dynamically stable planetary orbit is calculated for widely separated binaries. At bottom are close binaries. For a planet to bear life its orbit would have to fall completely within both the dynamically stable zone (*gray hatching*) and the thermally habitable zone (*colored hatching*). At the bottom the highly elliptical planetary orbit is dynamically stable, but it does not fall completely within the thermally habitable zone; the planet therefore could not harbor any life.

star could be detected by means of a long photographic exposure. Even the subtle methods used to detect spectroscopic binaries are not refined enough to find a planet, because the effect of a planet upon the motion of its parent star is so slight.

There is nonetheless other well-established evidence to support the contention that planets are common. In the first place, no sharp distinction can be drawn between binary or multiple stars and stars with planetary systems. According to Gerard P. Kuiper of the Yerkes Observatory, the mean distance of separation between the components of all binaries so far investigated is about 20 astronomical units. This is of the same order of magnitude as the distance between the sun and its major planets (Jupiter, Saturn, Uranus and Neptune). Kaj Aa. Strand at the U. S. Naval Observatory in Washington has studied small perturbations in the orbital motion of a star called 61 Cygni, which has a companion that is too faint to be directly observed. He has found that the mass of this unseen object is about a hundredth of that of the sun. This mass lies between that of stars and of Jupiter. It is therefore reasonable to believe that the masses of small stars in binary systems grade continuously down to the masses of planets. Since binaries are so common in our galaxy, it would seem that many stars that now appear to be alone actually possess planets.

Harold C. Urey of the Scripps Institution of Oceanography has found additional evidence for planets in a certain kind of meteorite. According to Urey, diamonds embedded in these objects show that they must at one time have been under high pressure in a body the size of the moon. Such moonlike objects, known as "prestellar nuclei," would enhance the formation of both stars and planets from a dust cloud. Any irregularity in the motions of a dust cloud should be expected to produce more than one such nucleus, and the formation of planets or multiple stars would follow as a normal consequence.

Finally, measurements of the angular momenta of many stars give every indication that planets exist outside the solar system. Otto Struve of the National Radio Astronomy Observatory has pointed out that main-sequence stars more massive than Type F5 usually rotate rapidly, but starting with this type the rotation of stars slows down abruptly. In other words, the average angular momentum per unit mass of the main-

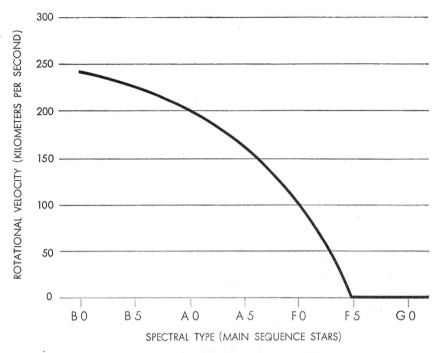

ROTATIONAL VELOCITY OF STARS on the main sequence is diagrammed here according to spectral type. Slow spin after Type F5 may indicate that planets are present.

sequence stars exhibits a conspicuous break at Type F5 [see illustration on this page]. The most reasonable explanation of this strange phenomenon is that unobservable planets have absorbed the angular momentum, just as Jupiter and other planets of the sun carry 98 per cent of the angular momentum of the solar system, leaving the sun with only 2 per cent and a comparatively long period (27 days) of rotation. If planets do indeed account for the slow spin of these otherwise sunlike stars, then planets appear just where life is most likely to flourish.

Thus it seems that intelligent life may be scattered throughout the Milky Way and the universe as a whole. In our immediate neighborhood, however, we may be its only representatives. The sun's nearest neighbor, Alpha Centauri, is only 4.3 light-years away. It is a triple system with two massive components (a G4 star and a K1) revolving around each other about 20 astronomical units apart; at a considerable distance is a small third star. The two larger bodies have highly eccentric orbits, and if there is a stable zone for a planet in this system, it is extremely hard to compute. Moreover, recent investigations indicate that the system may be much younger than the sun, so that higher forms of life might not have had time to evolve even if a habitable planet does exist in it.

Forty other stars are located within

five parsecs (16.7 light-years) of the sun. Only two—Epsilon Eridani (a K2 type) and Tau Ceti (G4)—seem to fulfill the conditions for the existence of advanced forms of life, and Epsilon Eridani may not be exactly on the main sequence. Tau Ceti is 10.8 light-years distant, has an apparent visual magnitude of 3.6, is located on the celestial sphere about 16 degrees south of its equator and appears above the horizon in the northern sky only in the winter.

Since intelligent life is probably not a rare phenomenon, and since at least one star in our vicinity meets the specifications of a life-fostering star, it may seem odd that we have had no visitors from other worlds. The idea would have drawn ridicule 20 years ago, but today it deserves consideration. There are, however, several reasons for believing that we have had no visitors from outer space. For one thing, the 10.8 light-years that separate us from Tau Ceti—astronomically a short distance—is an extremely long distance in terms of human experience. Even traveling at the speed of the artificial satellites that man has launched, space voyagers would need hundreds of thousands of years to traverse it. It is possible that organisms from Tau Ceti might have a far longer life-span than man, but this supposition invokes radical assumptions that cannot be supported by present knowledge. Furthermore, if a meeting is to occur,

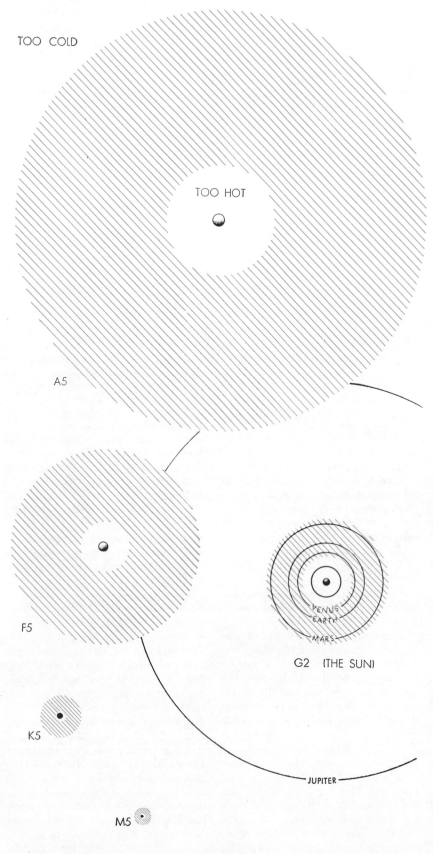

TOO COLD

TOO HOT

A5

F5

K5

M5

VENUS
EARTH
MARS

G2   (THE SUN)

JUPITER

THERMALLY HABITABLE ZONE of various types of star is here represented by hatched area around star. Here the A5 spectral type has largest zone, but star does not remain stable long enough for evolution to take place on a planet near it. Habitable zone of our sun (a G2 type) extends from orbit of Venus to orbit of Mars. Tiny M5 star has the smallest zone.

our high technological civilization would have to be contemporary with that created by intelligent organisms on other planets. The cultural evolution that brought mankind to its present technical competence began only a few centuries ago. And even if human civilization endures for hundreds of thousands of years, it would be a brief episode in the time-scale of biological evolution. Therefore the chance that advanced civilizations might be flourishing at the present time on the planets of one or two nearby stars is excessively small.

Some workers have concluded, however, that the chance is good enough —and certainly intriguing enough—to institute radio surveillance of signals originating outside the solar system. Guiseppe Cocconi and Philip Morrison of Cornell University have pointed out that the most favorable wavelength would be one close to but outside of the 21-centimeter line emitted by hydrogen in space, because in this region of the spectrum the galactic noise and the noise produced in the earth's atmosphere are at a minimum. Moreover, this important wavelength would be of as much interest to astronomers on another planet as to those on earth. Cocconi and Morrison have urged that radiation picked up by the 600-foot radio telescope, now being built by the Navy in West Virginia, be analyzed for the presence of signals. According to them, that telescope will be capable of detecting signals generated 10 light-years away by a technology no more advanced than our own. Meanwhile Frank D. Drake of the National Radio Astronomy Observatory, who makes a more generous estimate of the distance at which signals sent by intelligent organisms could be detected, is in charge of an actual project that is employing a smaller telescope to detect any radio signals transmitted by living beings in other "solar systems."

What kind of signals may we expect to receive or should we send out? Probably the most abstract and the most universal conception that any intelligent organisms anywhere would have devised is the sequence of cardinal numbers: 1, 2, 3, 4 and so on. The most likely signal would be a series of pulses indicating this sequence repeated at regular intervals. Such a signal may upon first consideration appear to be too simple for the sophisticated task of communicating with other beings far away among the stars. It would sound like baby talk. But after all interstellar communication is surely still in the baby-talk stage.

BIBLIOGRAPHIES

I FORMATION AND EARLY EVOLUTION OF THE EARTH

1. The Origin and Evolution of the Solar System

ORIGIN OF THE SOLAR SYSTEM. Edited by Robert Jastrow and A. G. W. Cameron. Academic Press, 1963.

EVOLUTION OF THE PROTOPLANETARY CLOUD AND FORMATION OF THE EARTH AND PLANETS. V.S. Safronov. Israel Program for Scientific Translations, Jerusalem, 1972.

SYMPOSIUM ON THE ORIGIN OF THE SOLAR SYSTEM. Edited by Hubert Reeves. Edition du Centre National de la Recherche Scientifique, Paris, 1972.

NUMERICAL MODELS OF THE PRIMITIVE SOLAR NEBULA. A. G. W. Cameron and M. R. Pine in *Icarus*, Vol. 18, No. 3, pages 377–406; March, 1973.

EARLY CHEMICAL HISTORY OF THE SOLAR SYSTEM. Lawrence Grossman and John W. Larimer in *Reviews of Geophysics and Space Physics*, Vol. 12, No. 1, pages 71–101; February, 1974.

2. The Origin of the Earth

THE PLANETS, THEIR ORIGIN AND DEVELOPMENT. Harold C. Urey. Yale University Press, 1952.

3. The Earth

SPECULATIONS ON THE EARTH'S THERMAL HISTORY. Francis Birch in *The Geological Society of America Bulletin*, Vol. 76, No. 2, pages 133–153; February, 1965.

HISTORY OF THE EARTH: AN INTRODUCTION TO HISTORICAL GEOLOGY. Bernhard Kummel. W. H. Freeman and Company, 1970.

EVOLUTION OF THE EARTH. Robert H. Dott, Jr., and Roger L. Batten. McGraw-Hill Book Company, 1971.

MODEL FOR THE EARLY HISTORY OF THE EARTH. S. P. Clark, Jr., K. K. Turekian and L. Grossman in *The Nature of the Solid Earth*, edited by Eugene C. Robertson. McGraw-Hill Book Company, 1972.

SEDIMENTARY CYCLING IN RELATION TO THE HISTORY OF THE CONTINENTS AND OCEANS. R. M. Garrels, F. T. MacKenzie and R. Siever in *The Nature of the Solid Earth*, edited by Eugene C. Robertson. McGraw-Hill Book Company, 1972.

FORMATION OF THE EARTH'S CORE. Don L. Anderson and Thomas C. Hanks in *Nature*, Vol. 237, No. 5355, pages 387–388; June 16, 1972.

EARTH. Frank Press and Raymond Siever. W. H. Freeman and Company, 1974.

II PREBIOTIC CHEMISTRY

4. The Chemistry of the Solar System

AN INTRODUCTION TO PLANETARY PHYSICS: THE TERRESTRIAL PLANETS. W. M. Kaula. John Wiley & Sons, Inc., 1968.

MOONS AND PLANETS: AN INTRODUCTION TO PLANETARY SCIENCE. William K. Hartmann. Bogden & Quigley, Inc., Publishers, 1972.

SPACE SCIENCE REVIEWS, Vol. 14; 1973.

5. The Origin of Life

The Origin of Life. A. I. Oparin. Dover Publications, Inc., 1953.

A Production of Amino Acids under Possible Primitive Earth Conditions. Stanley L. Miller in *Science*, Vol. 117, No. 3046, page 528; May 15, 1953.

Time's Arrow and Evolution. Harold F. Blum. Princeton University Press, 1951.

6. The Amateur Scientist: Experiments in Generating the Constituents of Living Matter from Inorganic Substances

The Origin of Life. Alexandr Ivanovich Oparin. Dover Publications, 1953.

The Origins of Prebiological Systems and of Their Molecular Matrices. Edited by Sidney W. Fox. Academic Press, 1965.

III   PROTOCELLS AND FOSSILS

7. The Oldest Fossils

Microorganisms from the Gunflint Chert. Elso S. Barghoorn and Stanley A. Tyler in *Science*, Vol. 147, No. 3658, pages 563–577; February 5, 1965.

Microorganisms Three Billion Years Old from the Precambrian of South Africa. Elso S. Barghoorn and J. William Schopf in *Science*, Vol. 152, No. 3723, pages 758–763; May 6, 1966.

Precambrian Micro-organisms and Evolutionary Events Prior to the Origin of Vascular Plants. J. William Schopf in *Biological Reviews*, Vol. 45, No. 3, pages 319–352; August, 1970.

Chemical Evolution and the Origin of Life: A Comprehensive Bibliography. Compiled by Martha W. West and Cyril Ponnamperuma in *Space Life Sciences*, Vol. 2, No. 2, pages 225–295; September, 1970.

8. Chemical Fossils

Chemical Evolution. M. Calvin in *Proceedings of the Royal Society*, Series A, Vol. 288, No. 1415, pages 441–466; November 30, 1965.

Occurrence of Isoprenoid Fatty Acids in the Green River Shale. J. N. Ramsay, James R. Maxwell, A. G. Douglas and Geoffrey Eglinton in *Science*, Vol. 153, No. 3740, pages 1133–1134; September 2, 1966.

Organic Pigments: Their Long-Term Fate. Max Blumer in *Science*, Vol. 149, No. 3685, pages 722–726; August 13, 1965.

IV   THE EARLY EVOLUTION OF LIFE

9. The Genetic Code: III

The Genetic Code, Vol. XXXI: 1966 Cold Spring Harbor Symposia on Quantitative Biology. Cold Spring Harbor Laboratory of Quantitative Biology, in press.

Molecular Biology of the Gene. James D. Watson. W. A. Benjamin, Inc., 1965.

RNA Codewords and Protein Synthesis, VII: On the General Nature of the RNA Code. M. Nirenberg, P. Leder, M. Bernfield, R. Brimacombe, J. Trupin, F. Rottman and C. O'Neal in *Proceedings of the National Academy of Sciences*, Vol. 53, No. 5, pages 1161–1168; May, 1965.

Studies on Polynucleotides, LVI: Further Syntheses, in Vitro, of Copolypetides Containing Two Amino Acids in Alternating Sequence Dependent upon DNA-like Polymers Containing Two Nucleotides in Alternating Sequence. D. S. Jones, S. Nishimura and H. G. Khorana in *Journal of Molecular Biology*, Vol. 16, No. 2, pages 454–472; April, 1966.

10. Symbiosis and Evolution

The Plastids: Their Chemistry, Structure, Growth and Inheritance. John T. O. Kirk and Richard A. E. Tilney-Bassett. W. H. Freeman and Company, 1967.

The Biogenesis of Mitochondria. D. B. Roodyn and D. Wilkie. Methuen & Co Ltd, 1968.

The Microbial World. Roger Y. Stanier, Michael Doudoroff and Edward A. Adelberg. Prentice-Hall, Inc., 1970.

Origin of Eukaryotic Cells. Lynn Margulis. Yale University Press, 1970.

The Oldest Fossils. Elso S. Barghoorn in *Scientific American*, Vol. 224, No. 5, pages 30–42; May, 1971.

11. Computer Analysis of Protein Evolution

PRINCIPLES OF ANIMAL TAXONOMY. George Gaylord Simpson. Columbia University Press, 1961.
COMPUTER AIDS TO PROTEIN SEQUENCE DETERMINATION. M. O. Dayhoff in *Journal of Theoretical Biology*, Vol. 8, No. 1, pages 97–112; January, 1965.
ATLAS OF PROTEIN SEQUENCE AND STRUCTURE, 1969. Margaret O. Dayhoff. National Biomedical Research Foundation, 1969.

V EXTRATERRESTRIAL LIFE

12. The Search for Life on Mars

SPECIAL VIKING ISSUE. *Science,* Vol. 193, No. 4255; August 27, 1976.
SPECIAL VIKING ISSUE. *Science,* Vol. 194, No. 4260; October 1, 1976.
SPECIAL VIKING ISSUE. *Science,* Vol. 194, No. 4271; December 17, 1976.
SPECIAL VIKING ISSUE. *Journal of Geophysical Research,* Vol. 82, No. 28; September 30, 1977.

13. Life Outside the Solar System

OCCURRENCE OF LIFE IN THE UNIVERSE. Su-Shu Huang in *The American Scientist*, Vol. 47, No. 3, pages 397–402; Autumn, 1959.
THE ORIGIN OF LIFE ON EARTH. A. I. Oparin. Oliver & Boyd, 1957.
THE PROBLEM OF LIFE IN THE UNIVERSE AND THE MODE OF STAR FORMATION. Su-Shu Huang in *Publications of the Astronomical Society of the Pacific*, Vol. 71, No. 422, pages 421–424; October, 1959.
STELLAR EVOLUTION, AN EXPLORATION FROM THE OBSERVATORY. Otto Struve. Princeton University Press, 1950.

INDEX